T0138388

Loving Faster than Light

Loving Faster than Light

Romance and Readers in Einstein's Universe

KATY PRICE

The University of Chicago Press Chicago and London

KATY PRICE is a senior lecturer in English at Anglia Ruskin University, Cambridge, England.

The University of Chicago Press, Chicago 60637
The University of Chicago Press, Ltd., London

Printed in the United States of America
21 20 19 18 17 16 15 14 13 12   1 2 3 4 5

ISBN-13: 978-0-226-68073-6 (cloth)
ISBN-10: 0-226-68073-8 (cloth)
ISBN-13: 978-0-226-68075-0 (e-book)
ISBN-10: 0-226-68075-4 (e-book)

Library of Congress Cataloging-in-Publication Data

Price, Katy.
Loving faster than light: romance and readers in Einstein's universe / Katy Price.
pages; cm
ISBN-13: 978-0-226-68073-6 (cloth: alkaline paper)
ISBN-10: 0-226-68073-8 (cloth: alkaline paper)
ISBN-13: 978-0-226-68075-0 (e-book)
ISBN-10: 0-226-68075-4 (e-book) 1. Relativity (Physics) in literature. 2. Relativity (Physics)—Press coverage. 3. Literature and science. 4. English literature—20th century—History and criticism. 5. Pulp literature, English—20th century—History and criticism. 6. Sayers, Dorothy L. (Dorothy Leigh), 1893–1957—Criticism and interpretation. 7. Eddington, Arthur Stanley, Sir, 1882–1944—Criticism and interpretation. 8. Empson, William, 1906–1984—Criticism and interpretation. I. Title.
PR478.S26P75 2012
820.9'36—dc23   2012007442

♾ This paper meets the requirements of ANSI/NISO Z39.48-1992 (Permanence of Paper).

They believe (though not all solemnly) that a love-affair is the fundamental means of understanding the world, or that the real purpose of building any system of knowledge is to understand love.

WILLIAM EMPSON, ON THE SEVENTEENTH-CENTURY METAPHYSICAL POETS

Contents

Acknowledgments

Thank you to all my teachers, especially Gillian Beer, John Haffenden, Michael Schmidt, Sally Shuttleworth, and Adam Strevens. My research has been made possible through the patience and curiosity of experts from across the disciplines. I would like to thank the following people for their support and conversation: Sylvia Adamson, Gavin Alexander, Matthew Bevis, Peter Bowler, Sarah Cain, Geoffrey Cantor, Anjan Chakravartty, Hasok Chang, David Clifford, John Constable, Robert Crawford, Gowan Dawson, Paul Day, Simon de Bourcier, Angelo di Cintio, Fern Elsdon Baker, Mogador Empson, Rebecca Empson, Elizabeth Falsey, Martha Fleming, Steven French, John Gardner, Jane Gregory, Jason Harding, Helen Haste, Rhodri Hayward, Linda Dalrymple Henderson, Peter Hingley, Simon Hodgkin, Iain Hood, John Holmes, Alex Houen, Jeff Hughes, Mark Hurn, Peter Jacobsen, Frank James, Alice Jenkins, Ludmilla Jordanova, Mara Kalnins, Melanie Keene, Candice Kent, the late Clive Kilmister, Jane A. Lewty, the late Peter Lipton, Simon Lock, Malcolm Longair, Jeff Mackowiack, Sophie Mayer, Willard McCarty, Felicity Mellor, Leon Mestel, Steve Miller, Simon Mitton, Paul Murdin, Jaume Navarro, Richard Noakes, Gavin Parkinson, Ian Patterson, the late Josh Phillips, Timothy Phillips, Adam Piette, Richard Price, Vanessa Price, Julia Reid, Christopher Ricks, Sharon Ruston, Simon Schaffer, Anne Secord, Jim Secord, Robert W. Smith, Matthew Stanley, Randall Stevenson, Rebecca Stott, the late Julia Swindells, Elizabeth Throesch, George Tiffin, Shafquat Towheed, John Tranter, Jon Turney, William Vanderburgh, Alice Waddicor, Eric White, Michael Whitworth, Rowlie

Wymer, and Tory Young. A big thank you to staff and students in the Department of English and Media at Anglia Ruskin University, the Department of History and Philosophy of Science at the University of Cambridge, and Homerton College, Cambridge, as well as to members of the Cambridge Science and Literature Reading Group and participants in a weekend course on Popular Physics and Modernist Literature at Madingley Hall in November 2004. My research has been further supported by the Department of English at the University of Sheffield and the Department of Science and Technology Studies at University College London. My research has greatly benefited from the expertise and generosity of staff at Cambridge University Library, the Whipple Library, the Wellcome Library, the British Library, the Wren Library, and the Houghton Library. I am very grateful for scholarships from the Arts and Humanities Research Council and for funding from King's College, Cambridge, the Centre for Research in the Arts, Social Sciences and Humanities, and the British Academy; for a Junior Research Fellowship at Homerton College; and for a year of research leave supported by Anglia Ruskin University and the Arts and Humanities Research Council and hosted by the Department of English and the Humanities Research Centre at the University of Sheffield. Karen Darling at the University of Chicago Press has been a magnificent editor, the copy editor Norma Sims Roche was incredibly helpful, and the anonymous reviewers of my manuscript were full of helpful and sensitive advice.

A Note on Money

All prices in this book are in pounds (£), shillings (s.), and pence (d.). There are twelve pence to the shilling, twenty shillings to the pound, and twenty-one shillings to the guinea.

In 1919, daily newspapers generally cost 1d. (the *Times* was 3d.), and weekly publications ranged from 1½d. for *Tit-bits* to 6d. for *Punch* or the *Athenaeum* and 1s. for *Nature*. A monthly magazine cost between 7d. and a shilling. These expenses may be compared with a tin of salmon at 7d., or beer at 8d. a pint. Books, from 2s. 6d. to 12s. and more, may be compared with tickets to a Royal Albert Hall Special Sunday Concert: 2s. 6d. for the gallery, 12s. for a reserved seat in the stalls, and £5 15s. for the grand tier.

An advertisement for Selfridges in the *Morning Post* for November 8, 1919, lists the weekly housekeeping budget for a family of four: a total of £3 4½d., including £1 4d. for rent, rates, and taxes. 5s. was set aside for amusements and the cinema, and the same amount for "Man's tobacco, papers, matches." This family would be living beyond the means of a clerk on an annual salary of £160.

Introduction

On November 7, 1919, newspaper readers in Britain awoke to a "revolution in science." But what did Einstein's new theory of the universe mean to public audiences confronted with headlines about space and time, light and gravitation? Announced to the public just four days before the first anniversary of the Armistice, relativity theory made headlines in Britain because Newtonian physics had apparently been overthrown by a German Jew. Two teams of British astronomers had conducted tests during the solar eclipse of May 1919, voyaging to far-flung imperial territories to measure the effect of the sun's gravitational field on light traveling to the earth from distant stars. Their calculations were announced six months later, at a joint meeting of the Royal Society of London and the Royal Astronomical Society on November 6, 1919, widely reported in the press. Journalistic celebration of "warped space" and "time as the fourth dimension" was countered by stern warnings about the folly of abandoning a reliable British ether for dubious continental metaphysics. One key feature of the new space and time stood out: almost nobody could understand or explain it.

This lack of an accessible, fixed interpretation heightened the theory's impact. Coming at a time of social, political, and economic upheaval, the mysterious new cosmology could be used to tell jokes and express fears about anything from cricket performance to Bolshevism. Public discussion of evolution theory during the nineteenth century had set a precedent for scientific claims to be explored from every angle and appropriated for diverging political, aesthetic, and moral ends.[1] Coverage of relativity in Britain quickly took on qualities of pastiche and parody as serious attempts to evaluate Einstein's theory jostled with self-aware commentary on media science.[2] Several papers carried substantial expository articles, including one by Einstein himself in the *Times*,[3] while jokes and satires linked relativity to everything from railway budgets to religion. The image of a befuddled newspaper reader attempting to explain Einstein's theory to his companions became a set piece in the popular press. The fact that readers couldn't escape from relativity became newsworthy in itself, and skepticism or resistance toward the new theory of space and time was coupled with more general doubts about an ever-expanding and diversifying world of print. "Commendation and condemnation of the theory have followed one another in bewildering sequence," wrote a reviewer for the journal *Science Progress* in 1924, "and even if we discard the journalistic type of article and opinions that are based on racial and nationalistic bias, we are still left with a vast array of articles, papers, and books on the subject."[4] What room could there be for yet more new arrivals in "the rather crowded relativistic temple"?

Articles and treatises aimed at all levels of readers continued to pour forth, but those keen to understand the theory and its implications were perpetually frustrated. Attempts at clarification tended to widen the gulf. "Only astronomers are affected" by the new discovery about gravitation, James Lockyer assured the historic meeting on November 6, 1919, a verdict reported in the *Daily Mail* and *New York Times*.[5] It was, a magazine article declared two years later, "hopeless to try to *picture* a 'space-time continuum'; to *imagine* space as 'boundless but finite'; to *see* the ends of straight lines as ultimately meeting."[6] Popular hunger for a fourth dimension that could be grasped was countered by science editors who wished to preserve scientific authority. This conflict was inflected through anxieties about labor unrest. "Nature knows no trade union, and is not guilty of strikes," readers of *Conquest: The Magazine of Popular Science* were informed in January 1920, in an article on hydroelectric power. In the same issue, Charles Davidson (one of the eclipse observers) observed that "Einstein's Law is not concerned with any difficulties which arise from our inability to conceive its physical meaning."[7] Seven years later, the situation had scarcely improved. "Incomprehension of the relativity theory is perhaps the most widespread human characteristic of the age," the influential literary journalist Arnold Bennett complained in 1927.[8] Investigating the "relativity affair" on behalf of "the great trade union of average intelligent persons," Bennett found that the latest books by philosopher Bertrand Russell and science journalist J. W. N. Sullivan fell short of delivering Einstein into the hands of "the ordinary intelligent person."

Multiple Meanings for Relativity

The story that emerges here is one of public frustration about the difficulty of Einstein's theory during the early 1920s, yielding by the 1930s to the sense that cosmic speculation had become a popular pastime. To illustrate this trajectory, I present excerpts from newspapers, magazines, and books that display a deliberate misapplication of the new cosmology to familiar experience. Ranging from a diamond thief who blames the fourth dimension for his crime in the *Daily Sketch* to a Quaker astronomer constructing an interplanetary love affair, every example collected here relishes the absurdity of the new space and time while using Einstein themes to express deeper concerns about life after the Great War (as World War I was called in Britain during this period). James Secord observes that the "texts of science have no meaning apart from what readers make out of them."[9] Some "texts of science" are more enduring than others, and it is the more ephemeral material from newspapers and magazines that most clearly demonstrates how even very abstract scientific ideas or language may be exploited as a rich and flexible cultural resource, resulting in multiple meanings that vary with audience and outlet.[10] Perfectly timed to meet its public one year after the Great War, the theory of relativity invoked a broad array of "other knowledge, other questions": the kind of extrascientific interest that Gillian Beer describes lying "latent" in any work, "waiting for the apt and inappropriate reader."[11] Or perhaps it is readers who lie in wait for discoveries, hungry for new theories and terms through which to reimagine their world.

The focus here is exclusively on British publications, with occasional glances across the Atlantic for comparison. As I have stated, the shift from frustration about Einstein to participation in a new universe was inflected through the class politics of the interwar years. This period saw the decline of the Liberal Party, the chief opponents of the Tories (Conservative Party) during the nineteenth century. Britain's last Liberal prime minister, David Lloyd George, had come to power when his predecessor Herbert Henry Asquith resigned in 1916 amid sharply divided views over the war. At this time the Liberals were in coalition with the Conservative Party. Lloyd George's government won the December 1918 election, notable for being the first in which women over thirty years of age and men without property were entitled to vote. But the troubled coalition broke up in 1922, when Andrew Bonar Law reclaimed power for the Conservatives after seventeen years in opposition. Meanwhile, the Labour Party was gaining ground, a reflection of increased support for socialism internationally. Socialism encompassed many different values, from the armed uprising of Bolsheviks led by Vladimir Lenin in October 1917 to the nonviolent "gradualism" espoused by Britain's Fabian Society (of which several figures mentioned in this book were members, including Oliver Lodge, H. G. Wells, and Arnold Bennett). As trade union activity and labor unrest culminated in the General Strike of May 1926, traditional British elites, threatened by the extension of the franchise, feared that Bolshevist sympathies might do irrevocable damage at home. In 1924 Ramsay Macdonald became Britain's first Labour prime minster, holding power for just a few months, but returning from 1929 to 1935.

The interwar years also saw newspaper owners gain increasing political influence. The *Daily Mail*, founded in 1896 by Alfred Harmsworth (Lord Northcliffe), played its part in the downfall of Lloyd George's coalition leadership. Rival press baron Max Aitken (Lord Beaverbrook) acquired the *Daily Express* in 1916 and was appointed the first ever minister for information (in charge of propaganda) in 1918. The Liberal *Daily News*, funded by Quaker businessman George Cadbury, switched its allegiance to Labour at the end of 1919. A reader between the wars was a voter, and reading about Einstein was caught up with concerns about how the newly enfranchised masses would determine Britain's future. Jokes about "scientific Bolshevism" in the press associated the Einstein revolution with social and political upheaval, and the question of access to Einstein's universe became freighted with the dynamics of class struggle. These political associations were compounded by the new theory's familiar name: the concept of "relativity" already had distinct philosophical connotations, and Einstein's theory was inevitably perceived through the Victorian debate about absolute versus relative values. A nineteenth-century relativist was a radical philosopher or scientist upholding secular values against the more conservative believer in absolutes.[12] This debate continued into the early twentieth century, resulting in the most famous dinner party gaffe in the history of physics: a beleaguered Archbishop of Canterbury, his ears full of Viscount Haldane on the liberating force of relativity, turns to Albert Einstein and ventures that the Jewish professor's work may make a great difference to "morale."[13] The association of relativity theory with relativism also gave it a lasting currency in commentaries on art and literature that sought to break with tradition.[14]

The concept of "relativity" took on a slightly different set of connotations in the context of mass entertainment. Tricks of perception were familiar from such devices as zoetropes and praxinoscopes as well as from the music hall. Populist outlets reclaimed "relativity" for this more tangible terrain, as when the clippings weekly *Tit-Bits* offered the minute scale of cheese mites and the mechanical ingenuity of a bicycling illusion as examples in its "Simple Explanation of a Very Difficult Subject," or *John O'London's Weekly* reproduced a column of gags from an American paper, headed "Explaining the Einstein Theory" and purporting to represent "Recent Researches along Broadway."[15] This wordplay implied that with the new cosmology, the great men of science had resorted to riddles and illusions, a view that was voiced more explicitly in American coverage.[16] In *Victorian Popularizers of Science*, Bernard Lightman has explored the ways in which showmen like John George Wood and John Henry Pepper capitalized on "the Victorians' hunger for spectacular visual images" in books and performances that demonstrated "the potential of science to attract vast, new audiences by incorporating visual spectacle."[17] The inability of Einstein's expositors to provide simple images encapsulating relativity made the new theory inherently "unpopular," yet at the same time helped to align it with technologies of illusion, especially the cinema. In addition, the invisible "fourth dimension" had preexisting occult and science fiction connotations.[18] Scientific writers on relativity theory had to negotiate these connotations, acknowledging the interest they held for readers while attempting to distance Einstein's theory from the romances of H. G. Wells or esoteric wanderings in the fourth dimension.

By 1937 the problem of access to Einstein's universe had been reversed. "The humblest is acquiring with facility, / A Universal-Complex sensibility," remarked W. H. Auden in his "Letter to Lord Byron."[19] In begrudging rhymes the poet laid the blame squarely with authors of best-selling popular physics books:

Impartial thought will give a proper status to
This interest in waterfalls and daisies,
Excessive love for the non-human faces,
That lives in hearts from Golders Green to Teddington;
It's all bound up with Einstein, Jeans, and Eddington.

Far from sparking a social revolution, enthusiasm for Einstein and his best-selling expositors had settled into a symbol of middle-class leisure activity and aspiration to higher culture. As with earlier press coverage, the experience of reading about the new physics was entwined with social and political identity. While Auden satirized a complacent affection for nebulae that masked a fear of Jews and "Reds," Nobel laureate J. J. Thomson, in his 1936 memoir, regretted the authorial tendency emerging in a thriving market for popular cosmology: "We have Einstein's space, de Sitter's space, expanding universes, contracting universes, vibrating universes, mysterious universes. In fact the pure mathematician may create universes just by writing down an equation, and indeed if he is an individualist he can have a universe of his own."[20] *The Mysterious Universe* (1930) by James Jeans and *The Expanding Universe* (1932) by Arthur Eddington were popular astronomy books by scientists well known for promoting idealist interpretations of the physical world (though Eddington's idealism is more prominent in his other books). These interpretations gave priority to human mental activity, and readers were thereby encouraged to believe that their own thoughts had a cosmic significance. Radical critics resented the bourgeois individualism associated with these works, while more conservative detractors implored popularizing scientists to stick to expert terrain and not stray into religion, philosophy, or politics.[21]

Readers of Einstein coverage and exposition were not only voters but also consumers, faced with increasing choices. By this time, newspapers obtained their principal revenue from advertising rather than sales (a shift known as the "Northcliffe revolution"). Relativity took its place among other commodities of the day. Readers who turned to a satirical piece about a time-traveling monk in the *Times* might also take in the adjacent "Fashion of To-day": "an evening cloak of Chinese blue and gold brocaded tissue, lined with dull tomato-coloured duvetyn and trimmed with deep fringes of monkey fur."[22] "Light Caught Bending," the *Daily Mail* revealed on November 7, 1919, beside an account of crowds swarming across London in search of lodgings for the Motor Show at Olympia. "£10,000,000 An Ounce": The *Daily Express* offered this calculation of the cost of light, according to the prices set by the gas and electric light companies. Magazines carried advertisements for correspondence courses in everything from engineering to journalism, inviting readers to improve their market worth. Authors of fiction and expositors alike recognized the experience of readers as consumers: they included commodities in their writing, from novels and whiskey to curative electric body belts, and marked the status of their protagonists through dress, dining habits, and leisure activities. In journalism, exposition, and fiction, the value of Einstein's relativity was thoroughly tested alongside contemporary phenomena such as telephone conversations and moving pictures, lady motorists and shorter dresses, psychoanalysis and modern marriage, electric light and margarine.

The whirl of new possibilities in the modern world was tempered by uncertainty and privation following the Great War. High prices and continued rationing meant that many families struggled to cover the basics, while the value of money was subject to postwar instability. Britain, along with other warring nations, had encouraged inflation in order to help finance its military effort. The gold standard was suspended in 1914, meaning that a pound note could not be converted into a gold sovereign. The British government had also borrowed heavily from the United States. Postwar increases in the bank rate helped to service this debt by driving up the value of the pound in an attempt to return to the gold standard. The relativity of measurement in Einstein's theory offered an obvious analogy for the slipperiness of national budgeting and international alliances. As a correspondent for the *Morning Post* warned, "If, under certain circumstances, a three foot rule can become the same thing as a six foot rule, then fourpence may really become ninepence."[23] The same paper proposed the existence of "a fourth dimension in international politics, . . . a power secret, occult, pervasive. It wields vast revenues and has plans which it conceals from its own dupes."[24] Prime Minister Lloyd George was held to be peculiarly susceptible to its influence, which was leading him "to cajole or trick this country into a negotiated peace with the Bolsheviks."

On a less explicit level, the strangeness and abstraction of relativity gave it associations with the unspeakable loss and dislocation experienced by soldiers and noncombatants. *Punch*, for instance, depicted a man who has tried to read Einstein's explanation of his theory in the *Times*: the commuter runs into difficulties on a railway platform, unable to interact smoothly with fellow passengers or his physical environment. Is this the effect of trying to envisage a fourth dimension or an attack of neurosis following shell shock?[25] Two artists entering the abandoned workshop of an experimental physicist accidentally reactivate a ray that has dispatched a man into the future. They glimpse the victim's body in a mute gesture of appeal, but are unable to pull him back into his proper temporal location before the ray sends him out of time once more. A pacifist astronomer proposes a four-dimensional love affair in which the lady must think of her lover for eight hours on end in the hope that their thoughts may overlap for a moment. An author of detective fiction holds this image in the background of stories in which shell-shocked men and newly independent women struggle to get on renewed terms with one another after the war. In these writings the mysterious qualities of Einstein's universe are found tacitly collaborating with separation and loss. As the new space and time resonated with changed class and sex relations and with new technologies of mass entertainment during the early 1920s, relativity became an apt symbol for an uncanny modern world in which exciting possibilities were matched by new risks and hazards.

Multiple Forms of Expertise

With a century's hindsight, how are we to interpret the diverse appropriations of Einstein themes for interwar audiences? Looking back, surely we can see how to disentangle physical theory from the politics, sentiments, and commercial fads of the early twentieth century. But there is much to be lost when science is set apart from its wider cultural appropriations. The jokes, complaints, and speculations provoked in the face of relativity's intense mathematical difficulty were not simply borrowing from the latest science story to address unrelated concerns. Those who produced and consumed this material were involved in renegotiating the relationship between "common experience" and powers of abstraction, helping to determine the scope and limits of science and asking how it related to other modes of knowledge and experience. Andrew Warwick, in his history of Cambridge mathematicians and physicists, has shown that the "disciples" of James Clerk Maxwell, who did not seize on Einstein's work as revolutionary, were not necessarily "slow" or mistaken: they had better solutions to the problems addressed by the special theory of relativity, according to their own research programs and expertise.[26] It is only with hindsight that acceptance of relativity theory becomes inevitable.

This principle of parallel expertise may be extended to mass-market appropriation of Einstein themes. "That the experiences of those who participated in the popular sphere differed in nature from those of professionals does not necessarily render those experiences superficial," contends Katharine Pandora in an essay challenging "the assumption of an incapacitated laity" in a different context, that of America before the Civil War.[27] What more famously "incapacitated" audiences could there be in the history of science than those confronted by the new physics of the twentieth century? Moving closer to the Einstein context, Graeme Gooday contends, in his work on the domestication of electricity, that "the popularization of technical specialisms should not be understood (simply) as an encounter between self-evidently 'expert' authorities and an inexpert or credulous laity."[28] In the case of electricity, "the typically divergent judgments among 'authorities'—even within the electrical community itself—positioned the laity as a tribunal of allegedly 'expert' judgments, being obliged to take their own informed decisions about which 'expert' to believe." This description is directly applicable to the Einstein sensation. As Michael Whitworth notes, "From its inception, relativity had caused dispute over whose authority counted."[29]

The following chapters show how producers of satire, news, poetry, and popular fiction brought their own special skills to bear on tensions between the abstract, general, or universal and the local, personal, or partisan. In media coverage, news values shape the handling of sources and selection of scientific detail. The *Daily Mail*, for example, presented relativity as a potentially misleading form of abstraction that needed bringing back within the scope of domestic experience. The *Times*, by contrast, called for enlightened readers to navigate a world shaped by relative values, leading less well-educated masses to make the right choices. Popular science magazines create an image of the intended reader and his or her abilities, but in the case of relativity, that image proved too restrictive for postwar readers, who were dissatisfied with the lines those magazines drew between abstract knowledge and everyday experience.[30] Authors of genre fiction (science fiction, crime and mystery, adventure, romance) specialize in knowing what their readers expect and in renegotiating the delivery of required plots and characters. Under these conditions the difficulty of relativity was subjected to genre devices that reasserted the validity of common experience while warning readers not to indulge too heavily in cosmic abstraction.

Given the range of expertise applied to Einstein themes, it is necessary to look for alternatives to the terms "lay," "nonexpert," or "nonspecialist" to describe consumers of scientific knowledge. I have adopted the term "mathematical innocents" to distinguish the baffled majority from the handful of adepts who had mastered relativity. Innocent to varying degrees, the writers who brought mass-market values to bear on the new space and time shared one key trait: they recognized the ability of popular audiences to take science stories seriously while at the same time finding science absurd. This duality, still plentiful in science coverage today, is a core skill possessed by those who consume science in culture: to simultaneously deride and support, to believe and disbelieve, to collectively roll our eyes while turning eagerly to hear more. "Encounters with science in the everyday world can be multifarious, miscellaneous, overlapping, partial, and contradictory—in fact, *un*disciplined," observes Katharine Pandora, echoing Gillian Beer's elaboration of "fleeting and discontinuous" encounters between literature and science.[31] My reading of Einstein in British popular culture blends an attention to agency on the part of audiences, drawn from the study of science communication, with an aliveness to ambiguity and ambivalence, learned through practicing literary criticism.

The difficulty of relativity theory, combined with its widespread prominence in mass culture, makes it an ideal testing ground for theories of science in culture. In a frequently cited article published in 1994, Roger Cooter and Stephen Pumfrey articulate an appeal that has been made many times over by scholars and practitioners of science communication: that we cease to view popularization as a one-way process in which knowledge is transferred from experts to public audiences, with an inevitable dilution or distortion along the way. They suggest the concept of "enrollment" as an alternative description of what happens when "representatives of learned science" appeal to an audience's interests in order to enlist support for scientific work: "When the lay audience accepts the appeal, it allows itself to become (indeed *makes* itself) part of a network of alliances which sustains that scientific enterprise. While the scientists have enrolled a public, so too have the public enrolled the scientists. According to its position and influence in the "network," the public alters the kind of science pursued in the future."[32] This description sounds promising, with the potential for equal exchange. In practice, enrollment can get messy. Each chapter here offers a complication: not invalidating the concept of enrollment, but showing how the excitement around relativity brought additional parties to the exchange between science and the public, helping to sustain some rather different networks than those prioritized by the "representatives of learned science." In elaborating these encounters I have opted to retain the terms "popular science" and "popular culture," despite the many reasons offered by historians for abandoning them.[33] My focus is always on how the popular is conceived and articulated in tension with elite culture, and retaining those terms is the most efficient way to describe that tension.

Three chapters on ephemeral genres of writing are followed by a further three on individual authors whose reputations have endured. Chapter 1, "Light Caught Bending," introduces the Einstein sensation through two spells of news coverage, triggered by the announcement of the eclipse observations in November 1919 and by Einstein's visit to Britain in June 1921. It analyzes a mix of satirical and serious coverage of relativity from national dailies—the *Daily Mail*, the *Times*, the *Daily Express*, and the *Daily News*—with supporting material from three weekly papers: the middle-class satirical *Punch*, the populist *Tit-Bits*, and the socialist *Time and Tide*. I show how the new cosmology was drawn into the service of news values at these different outlets, supporting political agendas and editorial policies while helping to consolidate each publication's relationship with its readers. Scientists used the press to gain support for divergent views of physics as a discipline, and journalists also used controversy in physics to raise questions about whose interests were being served—not just by science, but by government, new technology, and the mass media. Coverage of relativity reveals scientists and public audiences participating in a network of alliances that supported the news industry, but provided limited opportunities for the public to alter the kind of science pursued in the future.

In chapter 2, "Einstein for the Tired Business Man," I turn to popular science magazines, revealing the different strategies employed by editors and expositors as they attempted to rescue science from media sensation. Articles on relativity from two monthly magazines, *Conquest: The Magazine of Popular Science* (1919–1926), and *Discovery: A Monthly Journal of Popular Knowledge* (1920–1940), are compared. This comparison reveals the rather limited terms on which each magazine offered readers access to Einstein's universe. The class politics of this impasse are further explored through the handling of Einstein's theory in three very different publications: the quarterly general science journal *Science Progress*; the weekly paper of the Independent Labour Party, the *New Leader*, in which Bertrand Russell's *ABC of Relativity* was serialized in 1925; and the shamelessly populist monthly *Armchair Science* (1929–1940). These examples demonstrate that enrollment failed to happen through magazine coverage of relativity in Britain because the representatives of learned science were not able to acknowledge the interests of readers.

Chapter 3, "Cracks in the Cosmos," turns to Einstein-themed stories in pulp fiction magazines. Instead of having to struggle for authority against sensational news headlines, contributors to these magazines were free to draw the new space and time into the service of genre fiction conventions, such as eccentric professors, dangerous inventions, and supernatural intuition. As a result, mathematical innocents were depicted gaining access to Einstein's universe thanks to their vivid imagination, esoteric reading habits, and familiarity with new technologies of mass entertainment. But instead of celebrating the triumph of common experience, these stories all end by warning against the dangers of getting too closely involved with the new cosmology. This message was reiterated in romance plots featuring the new space and time, found across the pulp and glossy markets. My analysis reveals how these stories enrolled relativity themes in a half-formed protest against the incursion of intellectual advancement and mass media into everyday life. Relativity themes were used to hint at popular concern about the impact of communication technology on domestic life and social relations. But such concerns could not be directly voiced in publications that were intended to make a profit. Here the interests of readers were amply recognized, but the result was an affirmation of networks in support of mass-market fiction, not science.

Arthur Stanley Eddington (1882–1944) was particularly well placed to negotiate the impasse between relativity and common experience. Appointed director of the Observatory at Cambridge in 1913, he was in charge of the 1919 eclipse expedition to Principe, off the west coast of Africa, while another team went to Sobral in Brazil. Relativity appealed to Eddington, a committed Quaker, not only because of its promise of new knowledge about gravitation, but also because of the opportunity it offered to promote internationalism in science during a time of conflict.[34] As secretary to the Royal Astronomical Society, Eddington had received news of Einstein's work in 1916, and he worked carefully to persuade a skeptical community of British mathematicians and physicists that this German theory of the universe deserved serious consideration.[35] His technical *Report on the Relativity Theory of Gravitation* (1918) was followed by the semi-popular *Space Time and Gravitation: An Outline of the General Relativity Theory* (1920), which revealed hints of a "whimsical" style that was already beginning to make scientists uncomfortable.[36] Later, more accessible expositions of relativity appeared in contexts where religion was explicitly on the agenda. "The Domain of Physical Science," a chapter Eddington contributed to Joseph Needham's *Science, Religion and Reality* (1925), rehearsed key arguments for Eddington's Gifford Lectures of 1926–1927, published as *The Nature of the Physical World* (1928).

Eddington was an avid consumer of popular fiction alongside the literary classics, and he manipulated the devices of storytelling in his expository lectures and radio talks, which were written up into best-selling books. In chapter 4, "A Lady on Neptune," I draw on Eddington's unpublished letters and journal entries, alongside preliminary exposition from just before and after the eclipse expeditions, to demonstrate Eddington's sensitivity to textures of everyday experience.[37] I show how he invited readers to participate in establishing the significance of new physical laws in a world where intimacy, communication, and measurement had been disrupted by conflict. Eddington enrolled the public hunger for a common-sense account of relativity in an idealist interpretation of physics in which the scope of physical law was constrained as he made room for spiritual experience. While Eddington did much to make the new physics "popular," inspiring several future astronomers, his expository work did not result in a straightforward enrollment of readers in support of science, nor did it give them a stake in determining avenues for future research. What, then, was the value of his popularizing work? To explore this question further, I turn to two of his most careful readers, an author of detective fiction and a poet with a popular science addiction.

In chapter 5, "A Freak Sort of Planet," I analyze three stories by Dorothy L. Sayers (1893–1957), whose avid reading of popular physics has received little attention.[38] Sayers conceived of her most famous creation, the aristocratic detective Lord Peter Wimsey, during 1919 while teaching in Normandy.[39] Her eleven Wimsey novels and twenty short stories, published from 1923 to 1939, have been widely recognized as a mirror of British society during the "long week-end" between the First and Second World Wars. Critics have begun to establish the literary complexity of her writings alongside those of authors in the modernist canon such as Virginia Woolf, Naomi Mitchison, and Elizabeth Bowen.[40] I show how Sayers made popular physics a symptom of the postwar male disorientation that had to be overcome in order to initiate successful modern marriage. The epistolary novel *The Documents in the Case* (1930) features Jack Munting, whose obsession with popular science while he hovers on the brink of matrimony allows Sayers to explore traits that were also under development in Wimsey. The bond between shell-shocked men is given a fourth-dimensional twist in "The Image in the Mirror" (1933). Both these stories provide a context for Wimsey's claim, in "Absolutely Elsewhere" (1934), to have traveled at the speed of light from London to a crime scene forty miles away. I explore Sayers's deployment of Einstein themes as a social phenomenon rooted in characters' reading habits, comparing these stories to examples from Rose Macaulay (1881–1958) and Aldous Huxley (1894–1963).

In chapter 6, "Talking to Mars," I turn to a figure whose approach to popular science and literature has acted as a guiding influence throughout this study, in the form of aliveness to multiple meaning coupled with a humanist insistence on contradiction as the most telling feature of any creative endeavor. William Empson (1906–1984) is the youngest author featured here, born twenty-seven years after Albert Einstein. Having taken a degree in mathematics from Cambridge University in 1928, Empson went on to pursue studies in English literature, which resulted in the idiosyncratic yet influential critical book *Seven Types of Ambiguity* (1930).[41] Between 1928 and 1935 he published just over forty poems, many of which engage with themes from popular science. Astronomy, entomology, molecular biology, and botany are intermingled with allusions to classical and contemporary literature, philosophy, film, science fiction, and mass culture. The resulting verses are often dense, translating their passions into a restless web of meaning that stretches from ancient Greek thought to H. G. Wells. In these youthful poems, many of which were composed in the heat of thwarted love affairs, Empson was consciously imitating the metaphysical poetry of Renaissance author John Donne (1572–1631), to whose writings he maintained a lifelong commitment. Four hundred years earlier Donne had been inspired by the "new philosophy" of Copernican astronomy. In a succession of critical essays, Empson insisted that the theme of inhabited worlds allowed Donne to portray lovers as defying church authority and terrestrial morality.[42] In his own metaphysical love poems, Empson tested out the universe of Einstein and Eddington to see what resources it might offer adventurous lovers in an age of relativity. I reconnect these dense verses to themes in mass media coverage and popularization of physics and astronomy, opening Empson's poetic strategies up for a new generation of readers who may find them daunting.[43] I demonstrate the depth of his engagement with relativity by showing how he appeared to predict a black hole in a poem crafted between 1928 and 1935. Presenting the poems as a site of encounter between popular and elite sources, I affirm Empson's relevance to recent developments in the study of modernist literature, in which a restricted canon of innovative and elitist writing has been displaced in pursuit of exchanges between literary tradition and the mass market.[44]

The uses of Eddington's expository writing by Sayers and Empson reveal the complexities of enrollment in practice. Eddington's impact as an expositor was due to his deep understanding of Einstein's work, coupled with a genuine wish to enable negotiation between abstract theory and common experience. The resulting stories and analogies set in four-dimensional space-time were ripe for "apt and inappropriate" reinterpretation, besides inviting criticism from those who resented the astronomer's excursions into philosophy and mystical religion. Where Eddington enrolled everyday experience and human interest in an idealist account of physical law, Sayers and Empson populated his version of Einstein's universe with dislocated lovers and bewildered readers facing new choices after the Great War. Operating under distinct genre conventions, the detective author and poet each accepted Eddington's appeal but shifted its terms away from idealism and toward the reconstruction of intimacy. Rather than entering a network of alliances in support of any scientific enterprise, their writings act in the service of those who consume popular science. Sayers's stories and Empson's poems ask what else a reader needs in life, alongside science, and they push expository analogies to their extremity to see what happens when they break. Rather than altering how science was pursued in the future, Sayers and Empson each asked questions about the value, uses, and limits of science as a resource for thinking about wider social problems.

A brief concluding chapter, "Dreaming the Future," presents a postscript to the story of Einstein in British popular culture, showing how the stages of relativity's enrollment in British popular culture were recapitulated in the alternative time theory of J. W. Dunne (1875–1949). The reception of Dunne's book *An Experiment with Time* (1927) distilled and amplified the trajectory I have described here, from resentment at being excluded from Einstein's universe to delight and consternation at participating in the twentieth century's strange new time and space. The example of Dunne's work suggests ways in which findings about the reception of relativity can be applied more broadly to other science themes of the period and to our understanding of science in culture today.

ONE

Light Caught Bending: Relativity in the Newspapers

It is a Monday evening in June 1921. A time machine has just materialized in a London university lecture hall, but nobody pays any attention to the four young occupants who have arrived from hundreds of millennia in the future. The assembled audience are too busy attempting to follow Einstein's lecture, in German, on his new theory of space and time. On returning to her home at Multi Mansions on the Pacific Coast of Earth, Miss Diana Cosmos recounts the adventure in a letter to her friend Daphne, residing at Comet Castle in the ZZ Mountains on Mars. "Jack's bought one of the new long-distance racers," Diana tells her friend, and they have just returned from "a trip along the Fourth Dimension" with Jim and Audrey: "We went backwards for a change (I'm so bored with the future, aren't you, with our house crumbling to bits, and the dear Earth getting icier and icier, and all of us gone away to live on Venus, which isn't half so romantic as Earth). We didn't stop at the 500,000 year limit, as usual, because Jack, who's been studying primitive history lately, wanted to have a look at the man who first put mankind on the track of the Fourth Dimension."[1] Diana had, she tells Daphne, thought that the originator was "a primordinate named Wells," but Jack has set her straight: "Nonsense, it was Einstein, a contemporary." The foursome have accordingly voyaged all the way back to 1921, "which Jack said was the Einstein year, although I didn't know it."

Thus begins a report on Einstein's lecture at King's College London on June 13, 1921. Printed in the weekly newspaper *Time and Tide*, this satirical depiction of aristocratic pursuits in the deep future affirms the publication's socialist and feminist values. The time travelers find themselves in "a gloomy dim shut-in place full of chairs and weird human creatures with the littlest heads," and Diana is particularly fascinated by the women attending Einstein's lecture. An unerring authority on historical detail, Jack points out that "if we could see their ankles we should observe that they were wearing mostly champagne stockings. They were supposed to wear blue stockings, apparently, but none ever did."[2] It was, Diana is informed by her all-knowing consort, "an age of great uppishness and rebellion in the sex." The visitors learn that these women "couldn't possibly grasp Relativity, as very few even of the males could," although "contemporary female opinion was favourable to Einstein's hair, and his eyes were thought to be extremely touching." Observing carefully, Audrey wonders whether "the older, sober-looking women" might not have "absorbed a trifle of the theory," but is told that "they were only the wives of leading men present, and had no doubt come for the outing, and not to get wrinkles over the higher mathematics." Threats of uppishness in the sex have apparently subsided after more than five hundred millennia: "Isn't it droll to think that everything's changed since that weird time, everything that is, except Jim and Audrey and Jack and me, and what *we* stand for. Lucky Relativity hasn't altered *that*, what?"

The author of this satire, C. Patrick Thompson, articulates the paper's aspiration toward a future in which women will be able to talk about more than clothes and hair styling, coupled with a recognition that Einstein's theory cannot help to realize that vision.[3] The ability to grasp higher mathematics without getting wrinkles, or travel in the time dimension, will not inevitably bring about equality of the sexes. There is a class resonance to "what we stand for," hinting that the wealth and leisure enabling this foursome to zoom around in space-time have been maintained through the imposition of severe limits on social "relativity." *Time and Tide* commenced publication in 1920, with the dual aim of advancing women's influence in public life and educating women readers who were newly entitled to vote.[4] The comments about champagne stockings and tiny, bone-enclosed heads carry a regret that the intellectual sensation of the year is only helping to confirm the old order, with the celebrated theory remaining out of the reach of ordinary men and all women.

Jokes about Einstein's theory played out the defining drama of British society during the 1920s: new legislation meant that women over thirty and men without property could now vote, but could they be trusted with Britain's future?[5] Newspapers carried the terms of this debate into every corner of daily life, producing different images of what it meant to be a reader, and hence a voter. Publications like *Time and Tide* pitched themselves in direct opposition to the papers owned by Alfred Harmsworth, founder of the *Daily Mail* (in 1896) and the *Daily Mirror* (1903). Harmsworth, who had begun his career as editor of *Bicycling News* on a salary of £10 a month, also owned the *Evening News* (from 1894) and the *Observer* (from 1905) and had finally acquired the *Times* in 1908.[6] He was as much a part of the news as its leading proprietor, and in 1905 he accepted a peerage from King Edward VII, assuming the title Lord Northcliffe amid accusations that he had bought himself a place in the House of Lords.[7] Narratives of press power often identify the founding of the *Daily Mail* in 1896 as a symbolic moment, inaugurating an age in which commercial values dominated over rational debate and politicians found themselves at the mercy of newspaper barons.[8] Column inches could make or break political careers, and proprietors now made their money from selling advertising space rather than shifting copies.[9] A challenge faced by historians of the press is to work out the extent to which Northcliffe and the other newspaper owners simply gave readers the kind of news they wanted, or additionally used commercial strategies to create audiences for what they had to offer.[10]

Relativity first made headlines in November 1919, a year and a half before Einstein's lecture at King's College London and a few days before the first anniversary of the Armistice. The years immediately following the Great War were full of uncertainty, with continued rationing, threats of Bolshevism creeping over from Russia, railwaymen on strike, and unemployment high among returning soldiers. The new theory of space and time, with its warped rays of light and fourth dimension, offered rich resources for expressing hopes and fears that extended well beyond the scope of physical law. The fact that scientists themselves seemed unable to agree on its significance added a vital note of comedy to the proceedings. Mixing up politics, science, gossip, crime, money, and fashion, the newspapers of November 1919 provide a way to peel back a century of accumulated science and culture, enabling us to experience the new cosmology's arrival firsthand. By comparing coverage across Northcliffe's two main papers, the *Times* and the *Daily Mail*, we can get a deeper sense of what Einstein's theory originally meant to those "weird human creatures with the littlest heads." Newspaper content allows us to track the multiple meanings of relativity as it moved beyond sites of elite authority and entered the offices, parlors, and railway carriages populated by newly enfranchised readers.

Arithmetic Extraordinary

Readers parting with a penny to obtain a copy of the *Daily Mail* on November 7, 1919, were presented with a world of extremes. "Light Caught Bending," ran the science headline, alongside the news that a giant turtle had been found strolling across the sands at Cleethorpes the day before.[11] Two British expeditions, sent out to observe a solar eclipse, had successfully shown that "rays of light from the stars were deflected or bent in their passage past the neighbourhood of the sun in consequence of the sun's gravitation." The results, announced at a joint meeting of the Royal and Royal Astronomical Societies, were said to yield a positive result for the "new theory of gravitation put forward by a German mathematician named Einstein." Meanwhile, a shrimper named Kirman had been assisted in capturing the five-by-three-foot reptile, which had promptly been auctioned at Grimsby fish pontoon for £5 5s. and transported to London in time for the mayoral banquets. In the capital, the annual Motor Show at Olympia had reached similarly outsized proportions. The capacity of the vast Exhibition Centre had been exceeded, and a car now took the place of cakes in the window of a former tea shop at Hammersmith. Visitors were swarming as far as Kent and Brighton in search of hotel rooms, and in West London people were prepared to pay £1 1s. for a "shake-down" in a bath, lounge, or laundry room. Two men from Manchester were even reported to be sharing a single room on the top floor of a West Kensington boarding house.

More solemn news appeared in the form of a letter from the king "To All My People," calling for two minutes of silence at eleven o'clock in four days' time to mark "the first anniversary of the Armistice, which stayed the world-wide carnage of the four preceding years and marked the victory of Right and Freedom." George R.I. called for "a complete suspension of all our normal activities," a coordinated interval of time during which "all work, all sound, and all locomotion should cease, so that, in perfect stillness, the thoughts of everyone may be concentrated on reverent remembrance of the Glorious Dead." In villages, church clocks would indicate the designated time for remembrance. In towns with multiple clocks, all chiming at different moments, the Mayor would use a siren, and in London a maroon (a firework device, sounding like a cannon) would be set off. At eleven o'clock all the trains would stop, and police authorities across Britain were making arrangements to halt the traffic. Life across Britain would be completely frozen for 120 seconds.

The Armistice had been signed twelve months previously, but milk

THE DAILY MAIL, FRIDAY, NOVEMBER 7, 1919.

DEARER MONEY ALL ROUND.

THE BANK AND THE WASTERS.

WINTER WORRIES.

U.S. DOCK STRIKE ENDS.

I.W.W. HERE.

ENVOY TO 'CONVERT' WORKERS.

STRIKE SPECIALISTS.

The I.W.W. (Industrial Workers of the World), an American organisation with a none too pleasant history, have begun a campaign in this country with a view to converting British workers to their own views of life.

Mr. George Hardy is among the I.W.W. delegates sent to England, and he has already addressed meetings which were attended by Mr. Tom Mann, Mr. Robert Smil-

TWO MINUTES' SILENCE.

THE KING AND NOV. 11.

"REVERENT REMEMBRANCE OF THE GLORIOUS DEAD."

We are officially informed that the King invites all his people to join him in a special celebration of the anniversary of the cessation of war as set forth in the following

LIGHT CAUGHT BENDING.

A DISCOVERY LIKE NEWTON'S.

British astronomers, who watched the total eclipse of the sun in May last claim that by photographs they have proved that light, like matter, is subject to gravitation and may be deflected or bent.

The matter was discussed at a joint meeting of the Royal Society and the Royal Astronomical Society at Burlington House yesterday evening. The discussion was opened by the Astronomer-Royal, Sir Frank Dyson.

Two eclipse expeditions were sent out-

7

TURTLE WALKS ASHORE.

IN TIME FOR MAYORAL BANQUETS.

A Cleethorpes (Lincs) shrimper named Kirman was busy on the beach early yesterday morning when he was astonished to see a giant turtle stroll across the sands. With help he captured it. It measured, roughly, 5ft. by 3ft., and weighed nearly 3cwt.

The turtle was taken in triumph to Grimsby fish pontoon and offered for sale. The auctioneer solemnly assured the onlookers that the find had been made beneath Ross Castle, where it had existed for 350 years.

MOTOR SHOW TO-DAY.

LONDON PACKED.

OLYMPIA TOO SMALL FOR ALL EXHIBITORS.

EMPTY SHOPS FOR STANDS.

London is full of strange cars. People from all over the country and the Continent have crowded to

1. *Daily Mail*, November 7, 1919. © Solo Syndication and the British Library Board.

and food prices remained high. On the letters page, *Mail* readers learned of a clerk on a salary of just over three pounds a week with two small children and a milk bill of up to sixteen shillings, who was left with about two and a half pounds a week to provide all other necessities.[12] Those with enough to eat might turn their attention to the nation's spiritual health, thought to be especially perilous in the North of England. Rev. F. B. Johnson, president of the Leeds Free Church Council, reported a "grave danger of Leeds lapsing into paganism."[13] The railway system was also in jeopardy. A conciliation between railway unions and managers had been reached when war broke out, but negotiations had broken down once again in September 1919, resulting in a nine-day strike.[14] A publicity campaign promoting labor interests had shifted press criticism away from strikers and toward government railway investment.[15] On November 8, the *Daily Mail* noted that successive revisions to the annual railway budget had reduced it to £45 million, less than half the figure given seven months previously. The "Arithmetic Extraordinary" involved in such an adjustment was, the paper declared, "as hard to understand as Einstein's Theory of Relativity, which it puzzled all the wise men of the Royal Society to explain on Thursday night."[16]

Einstein perplexities were spreading quickly beyond the confines of the Royal Society, giving rise to a new genre of conversation that was set to become a staple of social satire during the 1920s. As the *Mail* reported, "Men tried to explain to each other all about the Newtonian law, Einstein's space . . . and Euclid's straight line, but became more confusing than convincing."[17] One journalist had sought out the astronomer James Lockyer (son of Joseph Norman Lockyer, the founding editor of the science journal *Nature*), who reassured readers that, while the new discoveries were undoubtedly important, they did not affect "anything on this earth," nor did they "personally concern human beings—only astronomers are affected." Encouraged by Lockyer's friendly explanation that, due to curvature of light rays, "any object such as a star is not necessarily in the direction in which it appears to be," the *Mail* reporter suggested a comparison with "the optical delusion of the position of a shilling in a tumbler of water." Here was an image of relativity at work in the parlor, a cosmic force capable of making the family shilling appear to shift about. The *Mail*'s approach coincided nicely with the plot of a musical comedy running at the Garrick in London from November 12, 1919. "The Eclipse" features a dubious professor of astronomy who gives two characters special pills that will render them immune to the effects of an upcoming eclipse.[18] Unlike everyone else in the hotel, the professor and his associates will be able to remember what has happened during the hours of eclipse. Various precious items go missing, with romantic tangles along the way, but in the end the professor is revealed to be the culprit, and love wins the day.

Visiting 1919 from a twenty-first-century vantage point, how are we to judge the *Mail*'s displaced shilling? The image might be dismissed as a misleading illustration of relativity, since it mixes up refraction (in which a wave of light changes speed and direction as it crosses the boundary from one medium to another) with the effect of gravitation on starlight. But some leading scientists at the time did appeal to refraction of starlight by the sun's atmosphere as an alternative to Einstein's arcane mathematics. What counted as a correct interpretation was still being negotiated, in public, at this point. The *Mail* reporter and his readers had other concerns in mind, however. The value of money had been manipulated in order to finance war, and the cost of living had risen severely during the conflict. The familiar optical effect is invoked here in an attempt not to counter Einstein, but to assert the significance of everyday material concerns. As an explanation of Einstein's theory, the shilling in a tumbler may be misleading, but this domestication of relativity serves an additional function, using the science story to question the elaborate financial schemes perpetrated by economic authorities, and vice versa.

The headline "Light Caught Bending" playfully implies that scientists do to natural phenomena what newspapers do to prominent social figures: catch them at a disadvantage. The *Mail* cleverly used the science story of November 1919 to pastiche its own news values, translating relativity into an occasion for celebrating all the reasons why anybody would want to pick up a copy of Northcliffe's leading paper. While astronomers revealed the displacement of stars appearing next to the sun, *Mail* news hounds were charged with exposing the displacement of other precious symbols: a giant turtle from Cleethorpes to the banqueting table; a new car into a window previously occupied by cakes; £55 million from this year's railway budget. The elitism of a new theory that affected only astronomers was jokingly associated with budget maneuvers, suggesting that the pursuit of scientific abstraction ran counter to public interest. At the same time, the displacement of stars in the heavens was reduced to the level of a parlor trick that could have no effect on the family budget, making abstract theory look like a harmless diversion from everyday concerns. Invited to associate high science with those two extremes, 55 million absent pounds and an enduring shilling in a tumbler of water, readers of the *Daily Mail* were given ample opportunity to mistrust and be amused by politicians and scientists alike. Such coverage affirmed the audience's expertise and acumen as spectators of the world's variety rather than participants in any form of radical action that might change it.

The Past and the Future

A copy of the *Times* contained more than double the number of pages in the *Mail*, used smaller print, and carried fewer illustrated advertisements.[19] Where the *Mail* exposed an unending variety of spectacle and extremity, the *Times* took pride in furnishing readers with plausible opinion from every angle. The modern world was awash with different outlooks, and it was only by continuing to read the *Times* that anybody could hope to reach a reliable verdict on the important issues of the day. The fact that nobody seemed to know what Einstein's theory actually meant made it the perfect story in this respect, and the *Times* went relativity mad for several weeks. The old guard of British physics were given ample space to warn against too hasty an elimination of the ether; a Victorian radical philosopher celebrated the vindication of relativism; the prime minister's untrustworthy smile was given free reign through space-time on a ray of light; a time-traveling monk was found casting spells over the beleaguered Church of England. Surveying the literary impact of the "Einstein Upheaval," *Punch* hastened in early December to reassure readers that Northcliffe had no plans as yet to rename his chief paper *The Fourth Dimensions*.[20]

For three times the cover price of the *Mail* or its rival the *Daily Express*, readers of the *Times* were taken inside the workings of imperial Britain. Forces threatening to undermine her stability were revealed, but simply by reading this newspaper, the king's subjects could participate in the rediscovery of order and unity, whether that meant knowing what time to send a postcard to Paris, appreciating the new knowledge possessed by lady motorists, or demanding that dairy farmers get rid of their unproductive cows. The initial headline, "Revolution in Science—New Theory of the Universe—Newtonian Ideas Overthrown," will be familiar to anybody who has previously encountered the story of Einstein's relativity.[21] But what did those words actually taste like on the morning of November 7, 1919, as the representatives of cultural authority turned to page twelve in the middle of their breakfast? The new cosmology entered a world of rising costs and rival interests held in precarious balance. The bank rate was going up, but the *Times* reminded readers that it had already predicted and explained this increase, which would make borrowing more expensive but would help to hold financial speculation in check. The price of coal, the profits of coal owners, and the cost of production were all under scrutiny in a renewed controversy that was, according to the *Times*, revealing further weakness and indecision on the part of the government. A revised bill for the regulation of wages and settlement of industrial disputes showed ministers crawling toward a compromise between trade union interests and those of the state, and a House of Commons debate on the matter had been "dull in the sense that a beef-steak is dull, but very English and very nutritious."[22] Full details of the new air mail service between London and Paris were given, including collection times from various post offices in London. Those who could afford 2s. 6d. per ounce could hand a letter or a commercial paper over the counter at Charing Cross or Threadneedle Street by eleven o'clock in the morning and have it arrive in Paris the same day, saving up to sixteen hours on the time taken by boat. The world was getting faster and more expensive, with greater scope for diverging outlooks, making newspapers even more necessary as a guide to modern life.

At the previous evening's scientific meeting, the Astronomer Royal, Frank Dyson, had convinced his audience that the results obtained by the two eclipse expeditions were "definitive and conclusive." Deflection of light from stars appearing close to the sun during totality had been measured, and the amount of deflection accorded with Einstein rather than with Newton. Readers heard the confession of Sir Joseph Thomson, presiding over the meeting, that "no one had yet succeeded in stating in clear language what the theory of Einstein really was," along with his conclusion that nonetheless "our conceptions of the fabric of the universe must be fundamentally altered." Peter Chalmers Mitchell, science correspondent at the *Times*, took a moment to recall Sir Oliver Lodge's address to the Royal Institution, delivered a month before the eclipse expeditions had set out.[23] Lodge had doubted that any deflection would be observed, but had been confident that if deflection did take place, "it would follow the law of Newton and not that of Einstein." Readers of the *Times* on November 7 learned that, as the president pronounced Einstein right, Lodge had left the meeting.

With his silvery beard and renowned stage presence, Oliver Lodge (1851–1940) was the British ether personified. He is the oldest figure in this story of Einstein's reception in Britain, but his research in physics covered vital technological and theoretical advances of his day: he worked on wireless and telegraphy, as well as being the most prominent exponent of James Clerk Maxwell's electromagnetic theory, on which the successful development of these new technologies depended. Lodge was an industrious popularizer, bringing the luminiferous (light-bearing) ether to a wide audience through books, lectures, demonstrations, and radio talks. For him, as for many others during the late nineteenth and early twentieth centuries, the ether carried more than light, radio, and other waves on the electromagnetic spectrum: it was also the bearer of messages from the dead. In the context of Victorian physics there were plausible connections between physical and psychic research, and spiritualist interests were sometimes pursued alongside scientific investigation.[24] Lodge's youngest son Raymond was killed by shrapnel in 1915 while fighting in France, and his parents were comforted by communications with their son through a medium, reported by Lodge in his widely read (and sharply criticized) book *Raymond* (1916, 1922). Lodge remained convinced that Einstein's relativity need not entail the demise of ether, a position he set out in *Relativity: A Very Elementary Exposition* (1925).

After perusing the newspaper report of that historic meeting on November 7, 1919, Lodge hastily wired a response to the *Times* from his Cardiff address. Far from being buried in the letters page, it was presented as a news item in its own right, under the headline "The Ether of Space—Sir Oliver Lodge's Caution—To the Editor of The Times."[25] "Sir,—To avoid misunderstanding," his letter began, "permit me to explain that my having to leave the meeting, reported in your issue of to-day (Friday), was due to a long-standing engagement and a 6 o'clock train." Describing the eclipse result as "a great triumph for Einstein" and praising the "excellent report" in the *Times*, Lodge denied that he had "ventured on anything so serious as a prediction" earlier that year. Nevertheless, he continued, the measurements that appeared to support Einstein might be explained without recourse to "great and complicated generalizations concerning space and time," perhaps by an appeal to "the ponderability of light coupled with a definite effect of motion on the Newtonian constant of gravitation" or even by means of "etherial constants" in conjunction with "a kind of refraction." Confident that the "splendid result" would be accounted for "with reasonable simplicity, in terms of the ether of space," Lodge offered gentlemanly congratulations to "Professor Einstein and also the skilled and painstaking observers who have so admirably verified his striking and original prediction."

Less ready to congratulate the German mathematician was Hugh Newall, a pioneer of astrophysics who had worked with J. J. Thomson at the Cavendish Laboratory. Newall berated fellow astronomers for their neglect of optical refraction and its many possibilities.[26] Edmund Nevill, an expert on the moon, wrote from Eastbourne to suggest that meteorites close to the sun might cause the same amount of deflection predicted by Einstein. Lodge was seen defending the ether once again toward the end of November, before an eminent audience that included Sir Rider Haggard and former prime minister Arthur Balfour at Lord Glenconner's house on Queen Anne's Gate.[27] All the while Sir Joseph Larmor, the British physicist whose work was closest to Einstein's electrodynamic research, was biding his time. A brief write-up of a meeting at the Royal Society on November 20 indicated that he was working on a revised version of Einstein's theory, and in the new year he wrote from Cambridge to affirm his belief that the triumphant connection of gravitation to the other forces of nature could indeed be "translated out of the fugitive and elusive notions of relativity" if only those involved could be persuaded to set aside their love of "mathematical intricacies" and focus on physical reality.[28] Far from representing the blind fixations of a passing generation, the views of Lodge and Larmor carried weight not only with the public, but also within the British scientific establishment of the early twentieth century, its disapproval of Lodge's psychic interests notwithstanding.

From pro-relativity quarters, the *Times* included articles explaining the theory's background and principles, including a translation of Einstein's own explanation, in which he described it as "a house with two separate stories, the special relativity theory and the general theory of relativity."[29] Readers were offered a faithful summary of Arthur Eddington's address to the Royal Astronomical Society on December 12, with subheadings that captured the *Alice in Wonderland* spirit of Eddington's expository style: "Space and Time," "No Difference between Straight and Crooked," "Trade Unionism of Physics," "Illogical Prejudices." Printed in small type with a good deal of technical commentary, these articles made sense only to those who already had some idea of how relativity worked. To the uninitiated it must have felt like being trapped in a geometry lesson under the command of Lewis Carroll's Red Queen. Coverage in the *Times Educational Supplement* leaned in the opposing direction, articulating resistance to non-Euclidean geometry.[30]

The new theory of space and time meant very different things to different people, and its relationship to questions beyond the scope of physics was a prominent feature of public discussion. An editorial printed the day the story broke affirmed Einstein's discovery that "the dimensions of space are not absolute, but relative and shifting" and declared that this discovery would "overthrow the certainty of ages," requiring "a new philosophy . . . that will sweep away nearly all that has hitherto been accepted as the axiomatic basis of physical thought."[31] A subsequent piece warned that the external world could no longer be relied on as a source of standards. "Observational science has in fact led back to the purest subjective idealism," the columnist declared, "if without Berkeley's major premise, itself an abstraction of Aristotelian notions of infinity, to take it out of chaos."[32] The octogenarian radical historian Frederic Harrison wrote in to remind readers that Auguste Comte, the founder of positivism, had declared the ether imaginary and antiscientific as far back as 1835. Followers of Comte's philosophy of relativity had long been disparaged by scientific authorities for their commitment to "practical certainty and relative truth."[33] Did Einstein's theory vindicate the Victorian relativists, who had bravely eschewed the pursuit of "absolute certainty, objective truth of the infinite All"? In defense of absolutes came an enormously long letter from the Oxford realist philosopher Thomas Case, flourishing a distinction between *causa cognoscendi* and *causa fiendi* (readers of the *Times* could presumably cope with such terms on a Saturday morning). Case, who could always be relied on to fulminate against anything excessively modern, from trade unionism to women at university, insisted on an elementary distinction between relative and absolute space: "If I go in a train from Weymouth to Oxford, carrying my own relative space with me," he patiently explained, "the whole space containing those two places and the interval between them is unaffected by my journey, indifferent to the rest and movement of all the bodies on the way."[34] Without this absolute containing space, a train passenger could start from Weymouth (where Case lived), but might never arrive in Oxford (where he ruled Corpus Christi with a rod of iron). Clearly Einstein and his supporters had been catching the wrong kind of trains.

As attempts at popular explanation of relativity began to appear, with their peculiar geometry and dizzying velocities, relativity's implications were heralded in the *Times* as "a gross affront to common sense."[35] What counted as news here was the challenge itself, and the display of resources that would be mustered to meet it, rather than the likelihood of a German physicist bringing about the demise of sensible thought. Loss of certainty and impending chaos were things that the British excelled at coping with, a strength that would be poignantly demonstrated during the two minutes of silence that were advertised alongside the "new philosophy." Uniting "the whole people throughout these Islands, the Oversea Dominions, and, indeed, every part of the Empire," this act of remembrance promised to "touch the inmost feelings of the British race" with a "majestic simplicity, native to their character and their habits."[36] The masses, who had hitherto been spectators of state ceremony and ritual, would for once be able to participate, and true equality would be achieved through the feelings and memories shared by "great and humble, wise and unwise, men and women, age and childhood."

Such equality did not quite extend to the masses streaming into London for the Motor Show. While the 2s. 6d. entry charge would help to exclude idle visitors who might waste the time of car salesmen, the influence of garagemen and chauffeurs on the motor industry had perhaps been overlooked: "It would be a sound business move," the Motoring Correspondent mused, to set aside part of one day when they might obtain entrance for 1s.[37] The show had become *the* social event of November, and those with 5s. or 10s. to spare on parading themselves at Olympia could obtain a ticket for Tuesday or Thursday daytime, either side of Armistice Day. It was rumored that the king himself might attend. Exhibitors at the 1919 show faced "a distinct change in the *personnel* of car buyers," for the "extended practical experience with motors" gained by women during the war meant that "the feminine owner" might now be more interested in the engine and chassis than in color, finish, and upholstery. But were British manufacturers, with their high prices and remote delivery dates, able to compete with foreign imports? The lodging houses of West London were bursting with potential motorists awaiting an affordable, efficient, and easily maintained vehicle such as American manufacturers "had begun to provide in thousand upon thousand long before the war." Would the British motor industry show itself awake to this demand, "its size, its universality, and its promise of rich rewards"?

More efficient production and supply were essential to Britain's health and prosperity in the face of continued hardship one year after the end of the war. Milk was in danger of becoming "an almost impossible luxury for the poorer classes," while basic foods had continued to rise in price since the Armistice.[38] Beer was now 8d. a pint, compared with 3d. before the war, and a bottle of whisky had gone up from 4s. to 10s. 6d. But the present administration appeared capable only of financial subterfuge to conceal the extent of these difficulties. A clever pastiche of popular astronomy and science fiction, headed "The Past and the Future," used Einstein themes to expose the inadequacy of coalition economics. Readers were invited to "imagine an intelligent observer receding from the earth with the velocity of light," setting out from the House of Commons at the moment when the prime minister was listening to the chancellor's revised, more optimistic account of the nation's finances.[39] Such an observer "would carry through all eternity, towards the limitless recesses of space, the bland smile of Mr. Lloyd George." It was, the satirist conceded, "an appalling thought." But what if the observer were traveling faster than light? He might then witness "the events precedent to the smile, the colloquy which doubtless led to the revision of the speech, the reading of the first edition, which led to the colloquy, and many other episodes which would be illuminating to ordinary mortals who have to guess at the past from the present." Here the satirist paused to observe that "a popular novelist" might be inclined to pursue this reconstruction of events, tracing "the evolution of a great statesman from the primordial slime." Turning soberly away from that sensational prospect, he proceeded to identify a new "metaphysical moral" arising from the relativity of time: "The present moment may be past, if referred to one set of co-ordinates; future, if referred to another." His message, invoking H. G. Wells, was clear: did readers envisage Lloyd George, with his bland smile and curved budgets, leading Britain's future, or would they consign him to the primordial slime?

This brief satire, penned once again by Peter Chalmers Mitchell, incorporates relativity as a disorienting modern addendum to evolution theory. The rejection of Newtonian absolute time enables the author to elaborate on the alarming prospect, recognized by the Victorians, that humanity might just as easily go backward as forward along the path of development. The new century brought a deep sense of disorientation, of not knowing which way is backward or forward, a theme that was taken up in the mass-market fiction of the early 1920s (explored in chapter 3). Just before Einstein's visit to London in 1921, the *Times* reprised the connection between relativity and evolution in an editorial headed "It All

Depends." A commentator held that Darwinian "survival of the fittest" had made success the proof of merit, lending philosophical support to the reformer's "shallow optimism" and the "ruthlessness of the aggressor."[40] This shift had now culminated in the loss of absolute standards in space, time, and morality, giving license to "ambiguous politicians" and allowing concepts such as justice "no meaning except in relation to a particular country or party or person at a particular moment." Following the lead of Viscount Richard Haldane, a former Liberal MP and lord chancellor whose lengthy work *The Reign of Relativity* had just been published, this editorial went on to explain that the challenge of a world without external standards could be met through careful definition of "the spheres within which given principles possess absolute validity." The human mind needed its own form of "constitutional government," and "healthy minds, like healthy communities, should beware of allowing their constitutions to be wrecked by adverse circumstances or overridden by absolutisms of any kind." The poor and the impressionable were more at risk from "misconceived absolutes" in an age of relativity, and the editorial concluded by quoting Haldane's call for readers to educate not only themselves, but also "those who, by their numbers, are our masters when the ballot takes place."[41] In Haldane's philosophy, the *Times* found a way to translate relativity into an affirmation of its own essential service to the public. Where *Mail* readers were cast as spectators to the world's variety, readers of the *Times* were invited to draw conflicting outlooks into a coherent, stable picture. Relativity was a signal to those with a more expansive view to take the lead in setting the coordinates of Britain's evolutionary time machine.

The Shadow-World

The "revolution in science" was inevitably associated with revolutionary politics. Under the heading "Trade Unionism of Physics," Eddington announced in the *Times* that "the rule of the great trade union of matter was that the longest possible time must be taken over any job."[42] Beyond Northcliffe's empire, the Einstein story was used to defuse threats posed by the increasing strength of trade unions. Jokes about scientific Bolshevism in the popular press were a useful way to make labor unrest look like a passing fad, as diverting and absurd as curved space or the fourth dimension. "Nitchevo!" exclaimed the *Daily Express*, delighting in the overthrow of "that infallible dogmatist, our old friend Euclid" and concluding that "Nothing is certain now."[43] On the Monday following the eclipse announcement, readers were entertained with a poem celebrating schoolboys' liberation from the pain of geometry lessons:

Smith Minor will emit a grateful shout,
And devastate the prep-room's still decorum;
The Powers of Darkness flee in utter rout
Now Einstein has blown up Pons Asinorum;
This sage, this kind Physico-Bolshevist,
Proves Euclid a deflated dogmatist![44]

In a similar vein, the *Daily Sketch* surveyed the implications of Einstein's theory, "which everybody is talking about and nobody understands."[45] Headed "Novel Possibilities in the 'Fourth Dimension,'" this article suggested that traditional geometry lessons might now be refused "on conscientious grounds," while a thief in the dock could avail himself of a most "Up-to-Date Defence":

"Your angle of vision is distorted, me lord," he might say. "I appear to you a crooked man, but let me tell you my 'angle of divergence' is so great that I am as straight as you."

"But what made you steal the diamonds?"

"Ah, there, my lord, we have an error in the fourth dimension."

Einsteinian wit was, it seemed, available to all but honest employees. Invited to imagine "the plight of a man with £200 a year applying for a rise of salary to an employer who believes in Einstein," *Sketch* readers were assured that the applicant would be told to consider his salary in the light of the fourth dimension, whereupon he would find that £200 was really £400. Similarly, "the happy father of twins who sees instead a fourth-dimensional quartette will probably set out to look for Einstein with a hammer." The founder of relativity became a convenient target for that symbol of united workers, the hammer. For the purposes of Edward Hulton's conservative tabloid, the mathematically abstruse science story offered a light-hearted way to deflect potential Bolshevist sympathies on the part of readers who might well be struggling to pay the family's milk bill. Even the familiar joke about schoolboys and Euclid was given a conservative turn, linking such protests to the conscientious objection for which pacifists had been jailed during the war.

The only national daily not in the mood for fourth-dimensional gags toward the end of 1919 was the *Daily News*. While *Express* writers were busy practicing their Russian, and *Mail* staff went about exposing the relativistic railway budget, an editorial in the *Daily News* soberly regretted the widening gulf between "a little knot of experts" and the interests of the "average man."[46] Headed "Unpopular Science," this piece made it sound as though the principal aim of physics was to make everyday life impossible: "Long ago the solid earth disintegrated beneath our feet into atoms; then we learnt that the atom is not the bedrock for which the weary mind longs in a world of flux; and now we are assured that space itself is not what it seems but 'warped.'" While this columnist did not doubt the honesty and energy of "tireless searchers after truth," he questioned their tendency to "plunge into worlds which grow ever shadowier and more unreal to the untrained understanding." And he conjectured that even Mr. Wells, with all his gifts as a popular writer, would surely be hard pressed to "make 'warped space' intelligible or interesting." These embittered comments are tinged with political regrets. Under Alfred Gardiner's editorship, the *News* had been riven by the same tensions that ultimately resulted in the demise of the Liberal Party. Following Gardiner's resignation in September 1919, the paper that George Cadbury had used to rally support for Lloyd George throughout the war finally turned its allegiance to Labour.[47] The alienation of familiar physical experience by warped space and fractured atoms offered a metaphor for the impending disintegration of a ruling power that had allowed the Conservatives only three years in sole government since the turn of the century. Pausing to hear "the knell of 'popular' science sounding," this *Daily News* journalist used Einstein themes to mourn the decline of more than a solid physical world.

Comments about mathematical difficulty and scientific rejection of common sense were found across the national press, with the curious exception of the *Daily Express*. Northcliffe's main rival, Max Aitken, had acquired this failing paper during 1916 just before the end of his six years as MP for Ashton-under-Lyne.[48] Having arrived in London from Canada in 1910, Aitken had quickly established himself as an influential force in British politics, and in 1918 he was appointed the first ever minister of information (in charge of propaganda).[49] When he was offered a peerage by Lloyd George, against the king's wishes, in December 1916, the *Morning Post* suggested that he adopt the title Lord Bunty—after the lead character in *Bunty Pulls the Strings*.[50] Opting instead for a name with rugged Canadian associations, Lord Beaverbrook went on to transform the *Express* into a profitable concern, acquiring in addition the *London Evening Standard* in 1923.[51]

An initial report in the *Express* described the "roomful of learned but somewhat bewildered astronomers" who had attempted to come to grips with "the new theory that light can be weighed."[52] Readers were informed that several eminent scientists had confessed an "utter inability to explain the theory in language which could be understood by a layman." A column the following day began by blaming photography for having helped to reveal that space is warped, concluding with the Russian peasant's exclamation, "Nitchevo!" This appeared alongside reports about the wonders of modern cinema, the amazing things that could now be done with women's hair, and the "hundreds of millions of gallons of rye and Bourbon whisky" that might be mobilized in an "alcoholic invasion of Great Britain" if Prohibition continued in the United States. News values at the *Express* emphasized physical sensation, bringing the world within reach of readers' eyes, hands, and mouths. The story about weighing light, in conjunction with references to photography and cinema, made relativity an adjunct to visual technologies: something to be experienced first and explained second.

This strategy stands out in contrast to talk of shadows and mysterious forces throughout the British press and the failure to give relativity a physical, picturable form in subsequent magazine exposition (discussed in chapter 2). A set of cigarette cards from 1928 resorts to the image most widely associated with Einstein's theory: a moving train.[53] Ogden's "Marvels of Motion" series illustrates action and reaction (boxers), centrifugal force (the circus horse), and gravity (Newton's apple) alongside technological achievements such as the cinematograph, racing motorboat, motor bus, and airplane and natural marvels such as the whirlpool, whirlwind, volcano, and meteor. "Einstein's Theory" is number five in the series, illustrated by means of a train struck by lightning in two places simultaneously while in motion. Text on the back explains that an observer on the train and one standing next to the rails would disagree on whether the two flashes were simultaneous. On card number sixteen, "The Motion Picture," the mirror image of an identical train is shown moving across the cinematograph screen, a subtle tribute to the connection between visual technologies and theoretical physics that had been established several years earlier. Whereas the train belonged to expositors, cinema was used to reclaim ownership of disrupted time and space on behalf of mathematical innocents. During Einstein's visit to Britain in 1921, a *Daily Express* cinema correspondent pressed the revered mathematician into agreeing that "the shadow-world created by cinematography" could help to illuminate his "theories about the unreliability of Time and the fickleness of Space." Seeking a connection to familiar experience, this journalist offered the cinema spectator's ability to perceive a moving image in the rapid succession of stills as a way to comprehend the cosmic view that might be attained by "a spectator of this world fortunate to be entirely removed from it."[54] Headed "Cinemativity—Time and Space as Puppets of the Screen," the interview appeared in print around an action photograph of female boxers in Berlin, their sports gloves set off with lipstick, neatly bobbed hair, and fetching white caps.

The *Express* for November 8, 1919, included a full column on the "new light discovery," between an account of traffic on Charing Cross Road held up by a dog dressed as a kangaroo and mention of the magnificent banquets awaiting President Poincaré on his visit from Paris. Headed "Upsetting the Universe," this article featured a substantial contribution from Dr. Charles Davidson, one of the astronomers involved in the Sobral eclipse expedition.[55] Davidson explained that the aim of these experiments, "stated in non-technical language," was "to prove that light has weight in proportion to its mass, as matter has." But he warned that "it takes an appalling amount of light to weigh an ounce." Based on the rates charged by the gas and electric companies, an ounce of light would cost something like £10,000,000. "This," he observed, "points the moral of daylight saving. The sun showers down on us 160 tons of this valuable stuff every day, and yet we often neglect this free gift and prefer to pay £10,000,000 an ounce for a much inferior quality." The first Summer Time Act had been passed in Britain in May 1916 as a means of saving fuel and keeping in step with Germany.[56] Attempts to introduce daylight saving before the war had foundered in controversy. In a House of Lords debate, Lord Balfour had emphasized the absurdity of this proposed interference with time, using the example of twins born either side of the clocks being set back in October. The order of their birth would be reversed, with the younger one being registered as the elder, leading to problems with inheritance law.[57] While most of Britain settled into acceptance of summer time, dissent in farming communities endured.[58] Charles Davidson's reference to the moral of daylight saving in the *Express* implied that the results of weighing light might be difficult and controversial, but would be accepted eventually. The terms "free gift" and "inferior quality" would have brought relativity even closer to the experience of readers who were confronted with increasing amounts of advertising and consumer choice. In this context, the ounce of light analogy acknowledges curiosity on the part of readers about what vision of the future science might be trying to sell, implicitly recognizing their status as electors and consumers.

The rest of Davidson's article focused on scientific facts about light—as a wave motion in the ether and as electromagnetic energy—and gave a brief outline of the procedures involved in photographing stars during an eclipse. There was no mention here of warped space, variable measuring rods, or getting lost in time: *Express* readers were being invited to evaluate a new experience, rather than marvel at the bizarre. The costly ounce of light had been used previously, by Davidson's senior colleague Arthur Eddington, as a way of distinguishing between mass and weight in public talks prior to the eclipse expeditions (see chapters 2 and 4). Coupled with Davidson's role as a practical worker in astronomy, the image suited news values at the *Express* perfectly, emphasizing immediate physical quantities. The experimental test of Einstein's theory was brought within the range of everyday experience, but, like the gallons of rye and whisky locked up under Prohibition across the Atlantic, the theoretical side of relativity remained tantalizingly out of reach for any readers who might wish to become more closely acquainted with the new space and time.

As it was drawn into the service of each newspaper's relationship with its readers, relativity polarized consumers of the new cosmology into two species resembling those described by H. G. Wells in *The Time Machine*. Advanced readers, like the vulnerable Eloi overlords, might have the leisure to indulge in speculation about the abstract nature of time and space, but their contemplative existence was increasingly under threat from hungry hordes, akin to the Morlocks.[59] This simple dichotomy was complicated, however, by a tendency (as we have already seen) among the British scientific elite to look askance at excesses of mathematical abstraction. The meaning of relativity was held in a delicate web of affiliation that crossed between Eloi and Morlock readerships. This becomes clearer when we turn to "the Einstein Year" of 1921. The most socially elite experience of relativity may be reconstructed by borrowing Jack's time machine to alight at an exclusive dinner party held three days before Einstein's lecture at King's College London. The hungrier end of the spectrum is found in *Tit-Bits from all the interesting Books, Periodicals, and Newspapers of the World*, a weekly paper that was held to epitomize the degenerate modern news style.[60]

A Very Dangerous Thing

Friday, June 10, 1921. Anyone who has glanced at a newspaper recently will know that Albert Einstein is in England for a week, giving public lectures on his theory of relativity.[61] Once more the journalists are in a frenzy, chasing round for interviews and marveling at the crowds gathered to hear him speak (in his native language) about this incomprehensible new theory of the universe. Randall Davidson, the Archbishop of

Canterbury, is among the guests at Viscount Haldane's house on Queen Anne's Gate. Haldane has, by this point, "retreated to the summit" of Einstein's theory, "where he breathes the pure air of Relativity and looks down with compassion on his fellow creatures still floundering in the Serbonian bog of Newtonian and Galilean physics."[62] Primed (if still confused) by Haldane's tireless exposition of relative ethics and politics, Davidson feels he ought to broach the question of the new theory's impact on religion. The Church of England's venerable leader turns to the Jewish professor of physics at his side and initiates an exchange that has become legendary.[63] "Haldane tells us that your theory ought to make a great difference to our morale," he begins (other versions of the story have "morals"). "Do not believe a word of it," Einstein replies at once. "It makes no difference," he continues, aware that an uncomfortable silence has descended. "It is purely abstract science," he offers. We may imagine the guest of honor gazing briefly at his wife Elsa, the archbishop seeking refuge in a glass of Château Margaux 1899.[64] It is a delicate moment: Haldane's *Reign of Relativity*, published just a few weeks before Einstein's visit, expends four hundred pages on the history and implications of "relativity" as liberation from dogmatic thought, heralding the visitor's theory as a mathematical affirmation of relative values in conduct and governance. Somebody makes a joke about the relativity of wine, perhaps; the professor smiles kindly; the conversation moves on. It is neither the first nor the last occasion on which Einstein will assert the irrelevance of his theory to human affairs.

Aged forty-two, Einstein is younger than almost everybody else in the room.[65] Haldane's sister Lady Elizabeth, who often takes time out from her own writing and social work to cohost these illustrious occasions, will celebrate her sixtieth birthday next year. Apart from Frau Einstein and Haldane's nephew Graeme, who is studying physics at Cambridge, the only other diners under sixty are three astronomers who are already secure converts to Einstein's theory.[66] The Astronomer Royal, Sir Frank Dyson, is in his early fifties, while the man whose name was all over the papers for proving Einstein right, Arthur Eddington, is not quite forty. Erwin Freundlich of the newly created Einstein Institute in Potsdam is a cub at thirty-six, half the archbishop's age. The company of scientific lions is augmented by Graeme's other uncle, the eminent physiologist John Scott Haldane, whose son J. B. S. Haldane is just embarking on a career that will include pioneering genetics, sexual scandal, and communism. Across the table the mathematical philosopher Alfred North Whitehead keeps a watchful eye on the honored professor, resisting the urge to trace calculations for his own version of relativity on the table napkin.

Tomorrow Whitehead will spend two hours trying to convince Einstein and Freundlich that they have got space and time wrong, driving the visiting celebrity to bed for an afternoon nap before his interview with the *Times*, in which he will confess that the finer points of Whitehead's scheme elude him.[67] Next in seniority to the archbishop is septuagenarian essayist and former House of Lords Librarian Edmund Gosse, whose two-volume biography of John Donne has helped to rekindle a fascination with metaphysical poetry among the new generation of poet-critics. But Gosse's own style of literary appreciation is the antithesis of all that "criticism" will soon come to stand for, and he is better known for an affecting account of his Plymouth Brethren upbringing and his father's resistance to Darwinian theory. His fellow ambassador of the arts tonight is Lady Frances Horner, who, still elegant at sixty-seven, modeled for the pre-Raphaelite painters Dante Gabriel Rossetti and Edward Burne-Jones when she was a young woman.[68] The calendar may register 1921, but the nineteenth century is not quite over yet.

The older generation will politely toast the conformity of light rays to the visitor's new law of gravitation, but they are by no means converted to his "relativity." J. J. Thomson is the same age as Haldane, but a good deal more skeptical about this new development in German physics. As president of the Royal Society when all the fuss broke out eighteen months earlier, Thomson found himself describing Einstein's theory as "one of the greatest achievements in the history of human thought."[69] But Thomson will go to his grave convinced that God and James Clerk Maxwell's electromagnetic ether are the only things a physicist—or anybody else—needs to believe in. His 1930 memoir includes a parody of relativity's elaborate time schemes, suggesting that a woman who has moved house between the birth of her two children "might appear to have had twins to an observer in a moving system, while to an observer in a system moving more rapidly the second child would appear to have been born before the first."[70] Thomson uses childbirth to assert physical experience over pure mathematics: "One wonders what the mother would have thought of relativity." Whereas Lord Balfour had appealed to aristocratic notions of birthright in the daylight saving debate, Thomson invokes maternal common sense to the same end: resisting the disruption of conventional order in space and time.

Einstein and his second wife are dining with eminently connected, influential representatives of British culture whose opinions and bloodlines are woven into the very fabric of British society. Will his theory be incorporated as another continental thread in the pattern? Or is something beginning to work loose—does the new cosmology augur the unraveling of traditional values? The world represented by Haldane's guests is far from invulnerable, as the recent carnage in Europe has all too clearly shown. Lady Horner's son Edward and her son-in-law Raymond Asquith were both killed in France. Raymond's father, Herbert Henry Asquith, nicknamed "the sledge-hammer" during his ascent to the premiership, was forced to resign as prime minister in December 1916, three months after Raymond had died of wounds from the Battle of the Somme.[71] Haldane was replaced as chancellor the previous year, in the wake of a press campaign against his German sympathies.[72] The current PM, David Lloyd George, is absent from the exclusive gathering to welcome Einstein, and his coalition Liberal government, formed by Asquith to maximize popular support during the war, is losing allies daily.[73] This will be the last Liberal government in Britain, and its crumbling position is not being helped by sharp attacks in Lord Northcliffe's papers.

Einstein's weeklong visit to Britain during the summer of 1921 caused a further flurry of press coverage on relativity themes, this time focused on the visitor's personality. The day before Haldane's dinner party, the *Daily News* informed readers of opposition to his "Jewish" mathematics from a group of German scientists.[74] Einstein's reputed affinity with children and his absent-minded "stage professor" image were emphasized: frequently drifting into a daze of theoretical speculation in the middle of university lectures, he had to be recalled to reality by "a chorus of coughs." In the *Express*, Einstein was represented as a patient and playful man, conceding—with "a twinkle in his eyes"—that "relatively speaking," his theory could be described in terms of the "cinemativity" proposed by his interviewer. Another *Express* correspondent had searched for Einstein a few days earlier as he docked from the transatlantic crossing at Liverpool. Applying the theory to the man, he had found that Einstein's "relativity . . . to his surroundings in space on board the *Celtic* to-day was not that of a fixed point. I had become a sort of ambient satellite with a devious orbit from lounge to state-room, then to the smoking-room, then to the promenade deck, and back to the lounge before I encountered him."[75] Noting the frequent newspaper comments on Einstein's "shock head of hair," *Punch* helpfully explained that the professor's hair "takes its own course through space and is not subject to gravitation."[76] The *Daily Graphic* had reported a "dark-skinned man" with "long black hair, a pair of intensely humorous dark eyes, and an ill-cut morning coat," while the *Daily Chronicle*'s professor was "dressed in well-cut black morning coat." Was this, *Punch* surmised, "another example of Relativity?"[77]

Even the *Daily News* was not immune to the professor's "little extra length of black hair" and "remarkable pair of eyes," commenting that except for these "he might be a prosperous South German man of business."[78] The *News* compared Einstein favorably to Bach, but complained that modern mathematical and musical achievements were too far from the reach of ordinary mortals: "It all depends on the exact size of the parietal lobe of the cerebrum whether one can follow Einstein in his mathematical treatment of the universe, with time as a fourth dimension, or really enjoy listening to the succession of cacophonous noises which Stravinsky calls music."[79] The Russian composer's woodwind symphony was judged "the last Strawinsky,"[80] while "a little knowledge of the Relativity Theory" was considered to be "a very dangerous thing."[81] There was, the *News* regretted, "nothing more pathetic than the sight of a man on a warm afternoon" attempting to explain how "the universe may be finite, but at the same time have no boundary." Enthusiasm for Einstein's relativity had become something of an embarrassment, reducing the average Englishman to a feminized state of confusion. Cosmological abstraction was, on the other hand, a welcome distraction from England's dismal performance against Australia in the Ashes. "'Relativity' Vies With Cricket," announced the *News*, noting the difficulty of finding a place in the crowded hall at King's College: "'As bad as Lord's on a Saturday'" was, apparently, "a remark heard more than once."[82] Meanwhile, *Punch* gleaned some helpful advice for cricket players: "'The earth does not emit a gravitational force which pulls the cricket ball down,' says an exponent of the Einstein theory in the *Observer*, 'it imposes a curvature on the surrounding space so that the path of the cricket ball appears curved, although it pursues the shortest course available to it.'"[83] Here, the pervasive mistrust of continental metaphysics is translated into distress at England's performance in the imperial sport.

Three months before Einstein's visit, an article on "relativity" in *Tit-Bits* reclaimed this term on behalf of the mathematically innocent masses. Headed "Why Nothing Stands Still—A Simple Explanation of a Very Difficult Subject," it did not even mention Einstein's name.[84] So much had been written about "the new theory of relativity," the writer observed, and yet it was doubtful whether "one person in a thousand has any idea of its meaning." But the essential point was easy enough to grasp. Words describing size, distance, or speed had no absolute meaning: "We speak of a huge gooseberry and of a small hill although the hill is obviously thousands of times larger than the gooseberry." The same principle applied to mice, houses, and (a favorite topic of the time, thanks to the first science film seen by a wide audience) cheese mites.[85] Turning from size to motion, *Tit-Bits* reminded readers that the earth's daily rotation on its axis gave an apparently stationary tree a speed of a thousand miles an hour, while "a train moving across Europe from East to West at 1,000 miles an hour" was, from this point of view, "really motionless." Viscount Haldane was cited as an authority for this "remarkable" conception, which was then compared to a "well-known music-hall turn" in which a cyclist rides his machine on "an endless belt which passes over rollers." In contrast to a train moving at a thousand miles an hour, the cycle act was based on a mechanical device that every reader could visualize: "When the rollers revolve, the path moves rapidly, and the cyclist, by dint of furious pedalling, is able to keep in one position upon it." Here, on the music hall stage, was relativity for all: "In relation to the stage and the auditorium he is standing still; in relation to the track he is moving at considerable speed."

A fortnight before the *Times* enlisted Haldane's interpretation of relativity in an appeal for healthy constitution, *Tit-Bits* invoked his authority in bringing the most abstruse scientific concept within the reach of anybody with pennies to spend on an evening's cinema or variety show. Beneath an article headed "Can Doctors Raise the Dead," the cheese mites and music hall cyclist were used to steer Einstein's theory away from abstract talk of trains, clocks, and planets and toward tangible experiences of ingenious entertainment. The vanquishing of mathematical abstraction here is aligned with the anxieties of Britain's scientific elite toward relativity, seen in Joseph Larmor's resistance to the "fugitive and elusive" qualities of relativity or in J. J. Thomson's distaste for mathematical invention. As with the *Daily Mail* shilling in a tumbler, it is possible to criticize the journalist for poor science: he or she is reasserting a pre-Einsteinian conception of relative motion. But this criticism would miss the function served for readers of *Tit-Bits*, just as a dismissal of Larmor, Lodge, and Thomson as scientifically defunct would miss the continuing significance of ether physics through the interwar years. Where the *Times* used Haldane's argument to align relativity with social responsibility, this "Simple Explanation of a Very Difficult Subject" diverts the theory into a celebration of mass entertainment, with consumers in the role of experts who know very well how to appreciate illusion and display. Just as astronomers know that bent light rays make the stars appear to shift about, music hall audiences know that the rollers make the cyclist appear to stand still, *Daily Mail* readers know that the shilling hasn't moved, and *Express* readers know that a cinema reel is really a rapid succession of stills.

What is the explanation for this coincidence between the opinions of elite physicists and material interests displayed in the popular press? The association of relativity with escape from reality could be mobilized according to different agendas. The *Daily News*, as we have seen, used it to mourn the passing of the Liberal Party as a cohesive force in British politics while declaring a renewed allegiance to common experience. The *Daily Mail* pursued abstraction so that readers didn't have to, challenging scientific theory and the Lloyd George administration at the same time. The *Times* invited readers to participate in abstraction, ensuring that its impact on reality went in the proper direction. The *Daily Express* and *Tit-Bits* countered abstraction with the mechanics of illusion making, allowing readers to feel that they had been behind the scenes of science while steering them away from giddy philosophizing. *Punch*, with its usual bite, gave voice to the hazards of not bringing readers back down to earth. Headed "Einsteinized," a piece from December 1919 described the excessive consumption of abstruse theory by mathematically innocent readers.[86] His mind "alive with science," a commuter attempts to navigate a busy train station:

> Arriving at a fixed point and accelerating my speed through a system of co-ordinates in a high state of motion, I followed the deviation of light rays to the end of the spectrum, and deposited my bag in the place for inert and heavy masses. Satisfied with my calculations I placed myself in a stable rotation, when unfortunately, while theorising, I collided with an immovable object.
>
> He opened up a whole string of new theories.
>
> I gave a warp into space and landed on my basal principle.

"I've given up science," the unfortunate victim concludes. The conversion of a fellow traveler's profanities into "new theories" and his own posterior into a "basal principle" parodies the desire of readers to improve their social standing through the consumption of irrelevant knowledge. But the skit also suggests that Einstein themes might be connected to a darker aspect of postwar British culture. The portrayal of a man becoming completely disoriented in the midst of everyday life is aligned with the symptoms of shell shock. "Einsteinized" allows the social comedy of a consumer's faith in newspapers to become tinged with the inexpressible injury and grief of war survivors.

The two spells of Einstein coverage in Britain, at the end of 1919 and during the summer of 1921, showed relativity themes fulfilling a number of intertwined purposes in the press: entertainment, commentary on the Liberal government, deflection of Bolshevism, affirmation of consumer identities, expression of postwar disorientation. As it was drawn into the service of distinct news values, the new theory of space and time took its place in British culture as a dislocating addendum to Darwin's theory of evolution, registering uncertainty as to what counted as a forward or backward development. The alignment of traditional values in British physics, in which physical conceptions were valued over mathematical abstraction, with the most lowbrow of news values, emphasizing physical sensation, made for a complex cultural politics that was set to harden into a simpler form as two apparently unrelated trends emerged. First, Einstein's relativity was gaining acceptance among scientists. Second, Labour interests were gaining ground as the Liberals declined. As the Eloi taste for abstraction was affirmed, Morlock hunger for physical conceptions grew. The flexibility pervading initial news coverage was lost as it became increasingly clear that participation in the new physics could only be on abstruse terms.

British newspaper readers might well have followed the moral of *Punch* and "given up science" if headlines about curved space and explanations involving high-speed trains were the sole medium through which Einstein's theory was presented to them. What the "Einsteinized" sketch leaves out is something that newspaper barons and scientific experts alike have almost no control over: conversation between readers. The *Daily News*, reporting the "pathetic" spectacle of a man on a warm afternoon attempting to explain finite but unbounded space to a companion, recognized the dangerous power of idle fourth-dimensional talk. Much of that conversation is unrecoverable, but whispers of it can be reconstructed from various sources. Chapters 3 and 5 show how authors of fiction depicted chatter about Einstein's theory as it unfolded out of the hearing of philosophical and scientific authorities. But it is in the Questions and Answers column of one particular popular science magazine that we come closest to hearing consumers of newspaper science speak in their own words. Chapter 2 explores editorial attempts to direct Einstein chatter along scientific lines through magazine expositions of Einstein aimed at countering the widespread misunderstanding of relativity that had been engendered through the pursuit of modern news values.

TWO

Einstein for the Tired Business Man: Exposition in Magazines

At the offices of *Conquest: The Magazine of Popular Science*, editor Percy W. Harris eyes his postbag uncertainly. It is January 1922, and the Einstein story has become the bane of his life. But a new article in this month's issue takes a different approach, and he hopes that the problem may now be resolved. The author of "What Is the Use of Einstein?" has adopted a sympathetic yet firm line, displaying an appreciation of "The Plain Man's Difficulties."[1] Harris allows himself a moment of glowing pride. He dusts the keys of his Noiseless typewriter in preparation for February's "Questions and Answers" column.[2] He can feel his bond with the Man in the Street deepening now, after two years at the helm of Britain's pioneering popular science magazine. He envisages men turning to one another in the morning train, smiles of satisfaction on their faces as *Conquest* is tucked away in preparation for the day's work. Higher subscription figures will secure the magazine's future, and perhaps he can even begin to dream of a new motorcar for Harris family excursions.

Not half an hour later, the editor of *Conquest* is slumped over his desk, torn between rage and hysterical laughter. The Einstein letters are madder than ever, filled with the most peculiar notions of things the professor may have overlooked. Harris is spurred by this voluminous misapprehension of mathematical physics into grim combat against popular delusion, against the widespread unscientific chatter represented in his postbag. The guiding light of *Conquest* must shine wherever there are newspapers, its writers countering "Fleet Street 'journalese'" with a keener sense of mankind's inventiveness and nature's munificence.

Published monthly from November 1919 to March 1926, *Conquest* promised to lead readers "into the laboratory of the scientist, the home of the inventor and the workshop of the manufacturer, where you may chat with them all informally and in plain, straightforward language."[3] For a shilling a month, anybody could gain access to "the why and the wherefore of the things that surround you in daily life," ranging from the "pure, nourishing and palatable" qualities of margarine to "huge dynamos lighting whole chains of cities."[4] The editor envisaged *Conquest* passing between "father, mother, sister and brother in every intelligent family," suggesting that the magazine might compete with the glossy fiction monthlies—some of which, like the *Strand* and *Pearson's*, also included nonfiction.[5] Each issue sported a vivid color painting on its cover, the words "science," "industry" and "invention" blazoned across. Balanced between entertainment and explanation, the articles were to "banish the boredom from science" with "no recourse to sensationalism, or the inaccuracy which so frequently characterises 'popular writing.'"

A regular column devoted to "Questions and Answers—Solutions of Readers' Difficulties" bore out the editorial promise of informal chat. This column was a crucial part of the magazine's monthly provision, assisting readers from Bodmin to Bangalore with a broad range of practical and theoretical inquiries. The column affords glimpses of the curiosity and technical expertise possessed by magazine readers almost a century ago. It also records instances of tension between readers and the gatekeepers of expertise. The issue for March 1921 gives a good idea of how this column worked.[6] A correspondent from Weymouth asked "how to make bichloride of mercury (calomel) from the metal mercury, not necessary to be pure?" The answer came that "bichloride of mercury is not 'calomel,' but corrosive sublimate." The correction was followed by instructions for producing mercurous chloride, or calomel (a traditional drug with uses ranging from laxative to fumigation), noting that "corrosive sublimate or mercuric chloride is more difficult to prepare from the metal, and seldom repays the trouble when done on the small scale." In the process of solving "Readers' Difficulties," the column often sharpened the distinction between readers of the "calomel" sort and the "bichloride of mercury" scale of science, industry, and invention. Another reader from Bangalore, who asked how to break down an electric current of 220 volts to 4 volts, was supplied with alternatives depending on "whether you merely want the low-pressure current for an occasional experiment or for some regular industrial purpose." Other questions came in a more piecemeal form that resembled the harvesting of news oddities displayed in *Tit-Bits*. All were given space, and the column's small type reflected the magazine's attempt to meet all inquiries. Having dealt at length with more sustained questions about the movement of the stars, the editor did his best to satisfy in a single paragraph the multifarious queries of a correspondent who wanted to know all about the young moon, diamond carats, Strads, mules, and Sirius, among other topics. A brisk response was given to each point, though the editor noted that matters of English usage and "the correct naming of vintages of port" fell "rather outside the scope of *Conquest*."

Strains in the relationship between questioning readers and Harris's team were evident from an article devoted to the more baffling contents of the *Conquest* postbag, printed in August 1921.[7] Sometimes, the editor mused, "I don't know" was the only answer that could be given to certain queries—not because the limits of knowledge had been reached, but "for the strange reason that many readers get angry when the faintest shadow of doubt is thrown on their pet beliefs." Two strange questions were, he reported, being "asked over and over again" by readers of *Conquest*: "Why does sunlight put the fire out?" and "If you put a kettle full of boiling water on your hand, why doesn't it burn you?" Some readers evidently found the tone adopted in the Questions and Answers column objectionable, and Harris noted that a "few have written suggesting that high-brow science sneers at obvious facts." The editor's temper was eased by quoting letters that were "as full of good common sense as they are of good-natured banter," and he offered these up as an example to his more tetchy correspondents. One reader from Belfast recalled a "fact well known up North," that a kettle containing a sufficiently boiled salt herring may be touched with impunity. There were some mysteries that might never yield to scientific conquest.

Advertisements played an equally important role in shaping the experience of popular science. Correspondence courses took up a significant proportion of the advertising space, offering instruction in journalism, short story writing, engineering, and management. These advertisements implied that every reader was also a potential writer or scientist and stressed the increased market value of those who had undergone specialist training. Sometimes, scientific images emphasized the difference in worth between trained and untrained readers: a set of scales weighing the employee against differently priced moneybags; a lump of charcoal contrasted with a diamond. In this context, even Fry's pure breakfast

2. *Conquest*, January 1920. Reproduced by kind permission of the Syndics of Cambridge University Library, L340:1.b.66.

cocoa and Nestlé's pure cow's milk exuded a sense of hierarchy, where only the best would suffice. Cover art and expository analogies reinforced this mentality. The magazine's third issue sported a particularly impressive cover illustration of a vast mountain in dark purple shades, looming over a smooth lake.[8] An unobtrusive red channel could be seen emerging from the mountainside, descending into a building whose arched windows lit up the whole page, the illumination reflected in the water and picked up in foliage surrounding the red roofs of humble alpine dwellings. Inside, the story of hydroelectric power explained that the "latent power" of the Norwegian snowfields is "no will-of-the-wisp, but a dependable agency," for "Nature knows no trade union, and is not guilty of strikes."[9] The "white coals" of these mountain heights "are always available when King Sol arrives 'to do his bit.'" The sovereign aesthetic of *Conquest*, clearly nervous about labor interests, shaped the first brave attempts to give Britain's popular magazine readers access to the new space and time. A major story in the same issue, dated January 1920, was "Weighing Light," by Charles Davidson.

What Is the Use of Einstein?

Relativity appeared at first to be an excellent topic for *Conquest* to address, proving the magazine's dedication to "the plain man" and his family. As the editorial for January 1920 observed, "newspaper discussions" of the new theory had "not been remarkable for their clarity," while Einstein's own article in the *Times* was "certainly not distinguished by that quality."[10] By way of remedy, Harris offered "an authoritative yet simply worded article on the British Eclipse Expedition and the theory it set out to prove, from the pen of no less an authority than Mr. Charles Davidson, F.R.A.S., in whose charge the expedition was placed." Filling nine pages, with thirteen illustrations, this was the editor's Christmas present to his readers: an overseas adventure story about the capture of starlight on photographic plates.[11] Davidson's article was filled with practical detail and contained some theoretical background, but he was reticent when it came to the fourth dimension and curvature of space. A more explicitly theoretical article, "Relativity of Time and Space," appeared just over a year later, followed by "What Is the Use of Einstein?" in January 1922 and "Some Difficulties in the Theory of Relativity" seven months after that.[12]

Davidson's account of the practical challenges faced by the astronomers at Sobral depicted high science in contact with the humblest circumstances of life in the Portuguese colony. A stark contrast between native inhabitants and visiting scientists haunts Davidson's narrative, evoking the imperialism of astronomical expeditions during the nineteenth and early twentieth centuries.[13] On arrival at Sobral, Davidson and his senior colleague Andrew Crommelin discovered that drought was driving most of the rural population to seek labor and shelter elsewhere. As guests of Colonel Vicente Saboya, the observers were able to take advantage of a private well serving his cotton factory, fitted with an additional pipe running to his house. Water was essential for developing the photographic plates obtained during the "300 precious seconds" of totality.[14] Before the night of the eclipse, an observing station was set up on the racecourse in front of the colonel's house. Masons were brought in to build piers on which the astronomical instruments could be mounted, and a painstaking process of calibration followed. The twenty-seven plates obtained during breaks in the clouds were developed on successive nights, in hot conditions that threatened to soften the gelatin film. With no ice available, the astronomers had recourse to "earthenware cooling pots in common use by the people," by means of which the temperature of the water was brought under 75°F.[15]

The image of astronomers improvising with common cooling pots during a season of drought may be contrasted with the experience of readers thirsty for details of the new theory. Davidson's article began with a discussion of light waves and the ether through which they were supposed to travel. He discussed "the famous Michelson-Morley experiment," designed to detect the earth's motion through that all-pervading medium, and used the example of two swimmers in a river to convey the expected result according to existing theory.[16] A swimmer who travels a set distance upstream and back downstream will take longer than a cross-stream swimmer covering the same distance, and light rays sent along the direction of the earth's motion through the ether and reflected back in a mirror were expected to take longer than those reflected back at right angles. Davidson then introduced the hypothesis that lengths contract along the line of their motion. This hypothesis, known as the FitzGerald contraction, had been devised to account for there being no discrepancy detected. He stepped neatly from the contraction of measuring rods to the realization that there can be no absolute measurement of space or time, reaching the nub of Einstein's theory only to declare it inaccessible: "Thus we arrive at the conception of a space in four relative dimensions instead of a space in the three absolute dimensions to which we have been so long accustomed." The astronomer declined to follow the argument any further "because it now becomes mathematically difficult, and physically obscure."[17] A familiar physical experience (swimming in a river) had been invoked to convey a technical point, but the outcome was a wider gulf between familiar and specialist knowledge.

Readers were given one last glimpse of theoretical interest before Davidson focused the rest of his article on predictions and experiments to test them. The relativity theory of gravitation entailed "a curvature of space," but he confessed that it was not easy to say what this really meant. "Einstein's Law," he explained, "is not concerned with any difficulties which arise from our inability to conceive its physical meaning. It is simply a mathematical expression of a law which is believed to fit into the observed facts." Narrating the imperial conquest of starlight,

Davidson's article demonstrates how expository strategies may establish a colonizing relationship between high science and familiar experience, where examples from everyday life end up reinforcing the inaccessibility of scientific developments rather than bringing technical knowledge within reach. But readers of *Conquest* were unquiet natives of the familiar world. In response to Davidson's swimmers, several readers wrote in with their own views on the matter. In the March 1920 Questions and Answers column, Harris reported that "two critics, having the courage of their convictions, proceeded to demonstrate that the journey up and down stream would be the shorter, despite Mr. Davidson's assertion to the contrary!"[18] A substantial portion of the column was taken up with a further elaboration of the relevant calculations, intended to "vindicate Mr. Davidson, and to disappoint our correspondents' hopes of upsetting the elusive theories of Professor Einstein."

The first article attempting to render those theories less elusive, "Relativity of Time and Space," was contributed by D. N. Mallik, a mathematics professor from Presidency College in Bengal. His Cambridge and Dublin degrees and his status as a fellow of the Royal Society of Edinburgh were advertised through an array of letters after his name.[19] Noting that "much of what has appeared on the subject in the newspapers, sometimes under sensational headlines, is somewhat wide of the mark," Mallik aimed "to give a simple account of the theory free from technicalities, if only for the purpose of removing misconceptions."[20] Emphasizing continuity with the past, he approached relativity in a different spirit than Davidson, urging that it was "possible to make a mental picture of the fundamental notions involved, when deprived of all recondite subtleties." His exposition featured what had rapidly become the customary equipment of relativity expositors: measuring rods, railway carriages, clocks, and lifts. Peter Galison's very readable book *Einstein's Clocks, Poincaré's Maps* (2003) reminds us that these items were closely related to the material circumstances informing Einstein's work. At the turn of the twentieth century, the distribution of time throughout cities, nations, and empires was vital to efficient business, safe running of the railways, and administrative control. Galison locates Einstein's theorizing alongside the many practical schemes for synchronizing clocks at the turn of the twentieth century, schemes with which he was closely involved through his family's electrotechnical business and his own work in the Bern patent office. But the practical Einstein was little in evidence in his own early attempt at making relativity accessible, in which the measuring rod, railway carriage, clocks, and a uniformly accelerating chest were introduced as expository devices.[21] By 1921 readers interested in the new time and space would have come to associate these objects with that gliding feeling where explanation and understanding part company. Mallik's dry and convoluted writing was unlikely to have helped, and readers of *Conquest* with hopes for more intimate contact with the fourth dimension would have gained little from the proposition that "time and directional extensions or geometrical co-ordinates . . . are correlated quantities."[22]

A year later, in January 1922, Harris displayed a sharper sense of his readers' needs by publishing "What Is the Use of Einstein?" This article was contributed by L. G. Brazier, whose name was unadorned with degree or fellowship titles, and it opened with an all-too-familiar scene: " 'This fuss and bother about Einstein is all very well,' says the 'Man in the Street,' looking out from behind his morning paper to his neighbour in tube, and tram, and 'bus, 'but what I say is "What is the use of it all?" ' "[23] Brazier appreciated that readers were both "mystified" and "resentful" at the amount of attention being given to "an abstruse theory" that apparently had "no remote connection" with the facts of anybody's existence. Reworking Davidson's original claim about the theory's lack of regard for physical meaning, he reassured readers that their inability to conceive of the fantastic curves and bends reported in newspaper headlines was perfectly natural. It was, he insisted, "hopeless to try to *picture* a 'space-time continuum'; to *imagine* space as 'boundless but finite'; to *see* the ends of straight lines as ultimately meeting." Instead, relativity had to be considered "an exercise in mathematical reasoning into realms beyond the capacity of our imaginations." Einstein's theory might one day open up a "glorious field of activity," if "the colossal stores of atomic energy" could be harnessed, but in the meantime it should be appreciated as knowledge for its own sake rather than for any "immediate utilisation."[24]

Far from stemming the tide of Einstein correspondence, Brazier's attempt at reassurance brought a "deluge" of letters demanding further information about relativity and offering various modifications of Einstein's theory.[25] Harris selected one example, printing it on the Questions and Answers page as a message to all: "G. H. E. (London, Q. 7).—*Question. 'Will you kindly give me a simple definition of the "Fourth Dimension"? I have been told by a friend that it is time-space; but when asked for a fuller explanation he could not give me one. Will it be possible to have a large article on the subject?'* "[26] A large article on the Fourth Dimension? The editor patiently explained that this topic had already been covered in depth: "*Answer.* Your friend's definition is certainly inadequate. Space in the ordinary acceptation has the three dimensions of length, breadth and depth. Einstein, for certain mathematical and physical reasons, has added 'time' as a fourth. To gain an idea of what this means you should read the three articles which have already appeared in *Conquest* on this subject." But readers of *Conquest* would not rest content with "certain mathematical and physical reasons" that nobody could explain. They wanted a fourth dimension that could be grasped. Devoting a further article to the increasingly bizarre contents of his postbag, Harris sighed at the alacrity with which correspondents appeared to have "picked up all sorts of odds and ends from the newspapers or even from uninstructed gossip."[27] They had, he reported, "exalted Einstein into a destroyer of common sense and a prophet of evil." He selected three letters in which readers chastised Einstein for his ignorance of geometrical progression and the refraction of light, and for his "elementary blunder of regarding Mercury as globular instead of spheroidal." Protesting that "a full investigation of the movement of the perihelion of Mercury on Einsteinian lines would prove uncommonly dry reading," the editor of *Conquest* effectively conceded defeat on the Einstein topic by providing a list of four books on relativity that he judged to be "within the reach of comparatively slender purses." Three of these were priced at 5s.: a translation of Erwin Freundlich's exposition of Einstein, a translation of Einstein's own lectures delivered at Princeton in 1921, and *Easy Lessons in Einstein* by American popularizer Edwin Slosson. More affordable at 2s. 6d. was *From Newton to Einstein* by another American author, Benjamin Harrow. Readers approaching either of the translated works would have found them tougher than Mallik's article. Turning to the American authors, they would have found more colorful analogies and illustrations bringing Einstein's universe to life, if not within reach. Both drew heavily on newspaper and magazine coverage. Slosson included Einstein's article from the *Times* as an appendix, the main text being compiled from a series of newspaper articles by the author that were themselves digests of various books and other articles on the topic, with slightly haphazard results.

Conquest's final attempt at explaining relativity demonstrates the vulnerability of magazine exposition, in contrast to the more secure relationship that book authors may establish with their readers. In August 1922 John Ambrose Fleming (inventor of the thermionic valve) set out to console readers for their exclusion from the edifice of physical theory in his article "Some Difficulties in the Theory of Relativity." In an inspired move, Fleming attempted to restore the metaphor of train travel to the realm of familiar experience by shifting its role in the service of relativity away from explanation and toward a more general statement about knowledge. Emphasizing the impossibility "of mentally picturing the nature of space in four dimensions," he offered a contrast between walking alone through an unknown country and traveling by train.[28]

When walking, he explained, "we must be able to *see* the way and follow the path step by step," but the train passenger does not need to be able to see the way in order to reach his destination.[29] Similarly, it was possible to arrive in Einstein's universe without being able to see the exact route: "The train is guided by the rails and transports us without the necessity for vision on our part. So the mathematical analysis conducts us to new truths though our powers of mental visualisation are strictly limited by previous experiences." Perhaps at the editor's request, Fleming included several sets of equations in his explanation of how relativity theory arrived at the deeper truth of "intervals" in the space-time continuum. These equations would have impressed on readers of *Conquest* a sense of the different class tickets available for travel on the Einstein train, some with a more satisfying view of the route than others. Fleming had shifted the analogy of train travel away from exposition and toward epistemology, but this only served to sharpen class distinctions around different levels of access to physical law.

With its nervous quotation marks around the "Man in the Street," *Conquest* betrayed uncertainty as to whether the street was an appropriate environment for scientific knowledge to frequent. Behind its confident article titles, from "The Conquest of Dust" to "The Conquest of Drought," the magazine was dogged by uncertainty as to whose interests ought to be served by this relentless extension of scientific dominion.[30] The problem of audience was one that faced all popular science magazines at the time, for "there were diverse potential readerships, none of which were big enough to generate a profit."[31] Restlessness under these conditions was apparent from the way *Conquest* kept changing its subtitle. Beginning as "a magazine of modern endeavour," it changed after a year to "the British magazine of popular science," dropped "British" eighteen months later, then in April 1924 became simply "the magazine of progress," and finally, in June 1925, "a magazine of progress, invention and discovery." In March 1926, announcing that the magazine would henceforth appear as *Modern Science (Incorporating Conquest)*, the editor wistfully looked back to its early days, when "conquest" had seemed an inevitable title, "symbolising the victorious advance of organised knowledge against ignorance and the disadvantages with which man faces his environment."[32] Reluctantly, the "more prosaic sound" of *Modern Science* was now being adopted, for the magazine needed "a less uncertain ring, a name to stamp it unmistakably for what it is, unless its spread is to be hindered and its contents to lie unsuspected between the covers." But *Modern Science* did not have much opportunity to spread unhindered, for the magazine was absorbed by its rival *Discovery* the following year. As

Peter Bowler observes, "the policy of writing down to the interests of the everyday reader had proved difficult to sustain."[33]

What the New Relativity Means

December 1919. Strolling past the newsstands of Oxford, Alexander Smith Russell eyes the lurid covers of *Conquest* with a mix of distaste and secret satisfaction. A professor of radiochemistry and fellow of Christ Church, he has spent the weeks of initial Einstein furor calmly overseeing the content for a rival publication. Launched in January 1920, *Discovery: A Monthly Popular Journal of Knowledge* was less colorful in its jackets and language, but had a clearer sense of how it would serve British scientists and their public. *Discovery* had close links to the driving figures in a campaign to redress the "neglect of science," a movement that emerged during the war in response to fears about Germany's greater expertise and efficiency.[34] A "journal" rather than a magazine, *Discovery* aimed to resolve the problem of public support for science in Britain by providing a channel through which specialists might communicate their findings to a wider audience.[35] Bowler explains that the journal was "very much an attempt by the scientific community to disseminate a particular vision of science's role in modern culture."[36] Far from sequestering science in its own special domain, the team behind *Discovery* intended to "take all knowledge as our province."[37] The range of its content may be judged by following a favorite title format through the 1920s. Readers were offered "New Light" on "Old Authors," on "a Neglected Century of British Sculpture," on "the Origin of Petroleum," on "the Ruins of Troy," and on "the Coal Measure Forests."[38] The first issue declared that expert contributors would write "not in the language of their specialism, but in that simpler tongue which we plain men do understand—namely, English." After an initial attempt to sell at sixpence, the cover price was raised in January 1921 to the inevitable shilling. Readers were effectively paying the price of admission to authoritative lectures in print form: "We hold that the specialist, when he has communicated his results to his fellow-workers in the ordinary way, should do the further work of making the same results plain to the ordinary man in books, pamphlets or articles easily understood."[39] For this "further work" to be completed, a willing audience was vital.

Articles in *Discovery* tended to include more technical detail, with fewer anecdotes or metaphors, than those in *Conquest*. Supported by an impressive array of learned societies, the journal encouraged appreciation of specialist knowledge for its own sake. Lists of suggested further reading at the end of each article conveyed a faith in readers' ability to become more closely involved in any topic that might capture their attention. Advertising also reflected the intended audience's involvement with amateur or professional science: optical lenses, telescopes, refractometers, and goniometers were promoted alongside books, journals, insurance policies, share companies, and charitable appeals. Contributors were often prominent figures, as a future-oriented series running through 1929 illustrates. Month by month, readers were presented with visions of "The Next Step" in various fields, including "The Future of Wireless Broadcasting" by Sir J. C. W. Reith (director general of the BBC), "The Archaeology of Tomorrow" by Osbert Crawford, "Aviation Spreads Its Wings" by Colonel the Master of Semphill (President of the Royal Aeronautical Society of Great Britain), and articles on physics and astronomy by Sir Oliver Lodge and Andrew Crommelin, respectively.[40] Reading *Discovery* was like receiving tuition from the leading British authorities in every conceivable field of knowledge.

Where *Conquest* writers tended to measure human achievements against those of nature, *Discovery* contributors instilled an appreciation of human ingenuity and natural phenomena alongside one another. By sharing professional concerns with the reader, they evoked a community of understanding around the significance of any given topic, from crystal structure to Roman trade societies. Einstein's relativity was difficult to stabilize along professional lines, however, because it was not immediately obvious to which community of British specialists it properly belonged. *Discovery* skirted the topic for as long as possible. In February 1920 the journal's second number included a contribution headed "Gravitation and Light" by the astronomer Harold Spencer Jones, chief assistant at the Royal Greenwich Observatory. This article began by acknowledging the widespread attention given to relativity "in the Daily Press and elsewhere" and noting that the "average person who has endeavoured to follow the discussion . . . has doubtless acquired but a very hazy notion of this theory."[41] The astronomer conceded that Einstein's theory "does, in fact, revolutionise our conceptions of space and time," but went on to state that "men of science are divided into two camps" on the question of its validity. Spencer Jones himself, an expert on the solar corona, held out in the "no" camp for several years. (A cautious man, he also dismissed the prospect of manned space travel beyond the moon in 1957, following the launch of Sputnik.)[42] His article for *Discovery* was not concerned with Einstein's theory itself, but with "the proof that a ray of light is bent when it passes near matter," an important discovery because "it is the first new thing which we have been able to learn about gravitation since the master mind of Isaac Newton enunciated the law of gravitation more than two hundred years ago." The main body of the article focused on technical and practical concerns pertaining to the eclipse measurements. He gave special attention to the excellent photographs of the solar corona that had been obtained (which had nothing to do with relativity) and asserted his belief that the results accorded with Newtonian theory. An editorial note in the next issue of *Discovery* was hardly more enthusiastic: "Einstein's theory of relativity has come to stay, so, however complex and incoherent it may appear to us at present, we must, we suppose, try to grasp what it is all about."[43] Readers were referred to articles in *Nature* and the *Times Educational Supplement* and told that an English translation of Einstein's own book was expected shortly. The translation was duly noted in September's "Books of the Month" column and praised in October's "Editorial Notes" for being "a tremendous help."[44] Readers would, A. S. Russell noted, require "a standard of education corresponding to that of a university matriculation examination, and . . . a fair amount of patience and force of will," but for their pains they would find "nothing in the mathematics or in the philosophical conceptions which need distress anybody unduly."

Dedicated readers of *Discovery* had to wait nine years for more than brief notices of books about relativity (though the review pages for August 1922 did include two excerpts from Eddington's Romanes Lecture, discussed in chapter 4).[45] An article in September 1924, "Measuring the Universe" by A. Vibert Douglas, aimed to show "how astronomers are actually making use of some of those brain-racking mathematical conceptions which we associate with the Einstein's Theory," but did not furnish an introduction to those conceptions.[46] An editorial for March 1929 suggested that relativity was merely a stage in some larger theory and promised an article on the latest developments. The following month saw "The Electron and Professor Eddington," in which Henry Francis Biggs related Eddington's quest for a fundamental quantity underlying all physical law.[47] June 1929 finally brought "What the New Relativity Means," also by Biggs, who located Einstein's achievement within his own specialist area of electromagnetism. This focus enabled the expositor to carry readers from Newtonian mechanics through Minkowski's space-time intervals and on to the most recent explorations of Arthur Eddington and the German mathematical physicist Hermann Weyl in just five pages, using no equations.[48] The fourth dimension and the curvature of space were handled with considerable tact, accompanied by apologetic asides regarding the limitations of language when it came to fully capturing such things. Biggs started by acknowledging that "we are all relativists by instinct," for we readily accept the results obtained in a terrestrial laboratory, treating them as if they had been obtained at rest even though we know that the earth is really whizzing through space.[49] The next step was to show how this "instinctive relativity," which was in keeping with Newtonian theory, came into conflict with modern electrical theory. A magnet in motion generates an electric field, and so a dispute will arise between a terrestrial observer traveling with a magnet that is at rest on earth and an observer from elsewhere in the solar system who insists "that the magnet is whizzing through space and producing something like a thousand volts per inch in the space between the poles."[50] Biggs went on to summarize the ways in which Einstein and the Dutch physicist H. A. Lorentz had sought to resolve this conflict. He then introduced Minkowski's absolute intervals between events and the "peculiar connexion between time and space" that resulted when measurement was liberated from its electromagnetic dependence on the observer's movements.[51]

All of this comprised "the first chapter of the history," leaving Biggs with only a page and a half for the second (on Einstein's general theory of relativity, incorporating gravitation) and the third (on subsequent developments).[52] The gravitation chapter occupied just two paragraphs because, as Biggs explained, the force of electricity "took no share in these doings, like the chairman of the town council who kicks off in the local football match and then retires to the most comfortable seat he can find." The third chapter was a challenge, but the theories of Arthur Eddington and Hermann Weyl deserved attention because they involved "cunning attempts to persuade electricity to join again in the fun." At this point Biggs confessed to finding Einstein's last paper (on general relativity) difficult to understand and harder still to explain. He concluded by wondering whether there might one day "arise some super-Einstein who will take what is best from the present Einstein, from Weyl, from Eddington, who will build with electrons, with quanta, with light, with gravitation, and weld it all into one magnificent whole."[53]

The editorial board of *Discovery* had wisely held off from dealing with relativity until they found an author who could stabilize it as a story unfolding within a specific community of expertise. Relating the Einstein story from the perspective of the electrical researcher and with the benefit of a decade's hindsight, Biggs presented relativity as a communal work whose final form was still open to negotiation. Any aspirations or anxieties that individuals might have when contemplating Einstein's universe could be assuaged by the prospect of specialist workers trying to get all the forces of nature to participate in a cosmic game of municipal football. The gulf in experience between football and electromagnetic research not only injected humor and humility, but also placed limits on the extent to which the analogy could assume a life of its own, thereby avoiding the unfortunate implications that attended Fleming's train journey. Questions of hierarchy or status—for example, whether it was appropriate for the town council chairman to get involved in a muddy game—became irrelevant. Instead, a straightforward relationship between players and spectators was established, with readers naturally fulfilling the second role. Whereas *Conquest* repeatedly informed readers that the experts could not furnish the "Man in the Street" with a picture of Einstein's universe, *Discovery* circumvented this troubling division by inviting readers to watch the community of experts fitting their distinct pieces of the picture together. There was no need for anybody to be grasping the fourth dimension here.

By this time the "monthly popular journal of knowledge" had absorbed its competitor, but *Discovery* still needed the financial support of academic and professional institutions. As Peter Bowler has pointed out, the journal never made a profit.[54] Coverage of relativity in the monthly popular science outlets was intended to counter the deleterious effect of sensational news headlines, but in setting out to amend the relationship between science and the reading public, popular science magazine editors and contributors found the Einstein story hardening into an impasse between the elite owners of scientific knowledge and imaginative work done by mathematically innocent readers. The realms of abstraction were deemed out of bounds to inhabitants of the common physical world, and relativity became a rarefied commodity beyond the reach of those "slender purses" Harris and his writers were aiming to satisfy. The brave attempts at relativity exposition in *Conquest* were thwarted by this dynamic, while *Discovery* evaded the problem without reaching a resolution.

The Great Trade Union of Average Intelligent Persons

The labor dynamics of relativity exposition were made explicit on humorous terms in 1927 by Arnold Bennett in his weekly book review column, "Books and Persons," for the *Evening Standard*. A popular novelist and literary journalist, Bennett found it strange that, "after all these years," he understood "little more of the relativity theory than a clever hall-porter."[55] And he was not alone. "Incomprehension of the relativity theory is," he observed, "perhaps the most widespread human characteristic of the age." Contemplating the notion that "a theory which re-states the physical universe must remain for ever a perfect mystery to the plain man," he retorted, "Plain men are not so plain as all that." One essay on Einstein's theory, in the collection *Aspects of Science* (1926) by science journalist J. W. N. Sullivan, had, Bennett declared, done "some useful damage to the walls of my ignorance." Hopeful of a sales boost, several publishers followed up by sending copies of their latest relativity books for his perusal. Bennett opened his column for April 21 by noting that he had received "sympathetic letters from readers" in response to his confessed incomprehension of Einstein.[56] This had spurred him to action: "I went about among my brilliant friends asking: 'Do you understand relativity?' 'Do *you*?' 'or you?' Nobody could say yes. On behalf of the great trade union of average intelligent persons, I felt humiliated. I decided that I would go into the relativity affair myself. I now report to the trade union." Under the heading "Einstein for the Tired Business Man," Bennett weighed the merits of Bertrand Russell's *ABC of Relativity* (1925), J. W. N. Sullivan's *Three Men Discuss Relativity* (1925), and *From Kant to Einstein* (1926), by one Hervey de Montmorency. The most successful book would, he decided, enable readers to answer the question, "Now what *is* Einstein's theory?" Russell's exposition was pronounced a failure on this count. Sullivan did succeed in giving the reader a sense of what relativity was about, but had opted—perhaps wisely—not to attempt a presentation of the theory itself.[57] In the end, the review nominated de Montmorency's slim volume as of most use to "the tired business giant determined to improve his mind of an evening."[58]

Confident in his duty to *Standard* readers, Bennett went beyond the task of evaluating the expository writing to challenge the theory itself. Presenting a list of "conclusions which I, the ordinary intelligent person, have reached about Einstein's Theory," he identified two mistaken formulations. Bennett suspected that the founder of relativity was "not expressing himself very happily" in the claim that "space is curved" and went into detail about how these two words could legitimately be combined: "A space may be confined within curved limits. A space may be occupied by bodies that move in curves. But surely space itself cannot be curved." Similarly, he felt that it was "merely playing with words to call 'time' a fourth dimension." These comments go beyond a criticism of the language used to popularize relativity, echoing the mistrust of mathematical abstraction that was displayed by British physical scientists responding to relativity: "Higher mathematics can postulate four dimensions, even forty dimensions, and can work out sums in them," he observed. "But my mind cannot picture more than three dimensions, and no mind can picture more than three."

There is something slightly peculiar about a trade union that numbers tired business giants among its membership. Although this review appeared less than a year after the General Strike of May 1926, Bennett's labor rhetoric would not have been taken too seriously by readers of the conservative *Evening Standard*. The phrase "great trade union" was in habitual light usage during this period, as seen in Arthur Eddington's lecture reported in the *Times* in December 1919, or in a later story of Bennett's featuring "the unofficial representative of the great trade union of husbands" being kissed in forgiveness of his enraged accusations and violent conduct.[59] But this does not mean that "Einstein for the Tired Business Man" was devoid of any political connotations. In conclusion, Bennett supposed that Einstein would one day be superseded, and that later on "the genius who has superseded him will be superseded by somebody else." This verdict echoed the spirit of gradual socialist reform advocated by the Fabian Society, which Bennett had been a member of for two years up until 1909, along with his friend H. G. Wells as well as George Bernard Shaw and Oliver Lodge. Bennett saw Einstein as one of a succession of inspired thinkers, and he confessed to having found himself "as excited by these brief and superficial studies as I ever was by an ode of Wordsworth's or a novel of Hardy's."[60] High praise indeed: Bennett was notoriously intolerant of difficult writing on the part of literary modernists, dismissing their convoluted prose in much harsher terms than those used for the Einstein expositors.[61] By placing relativity theory on a par with nineteenth-century literature, he afforded the new cosmology a cultural status that was subject to its being brought into line with certain convictions about realism and representation. In convening the great trade union of average intelligent readers, Bennett was calling not for radical action, but for renegotiation of the relationship between abstraction and common experience. He wanted the work done by readers to be awarded greater recognition in Britain's intellectual economy, a theme that resonates throughout the Books and Persons column.

Tensions between elite science and the mass of readers pre-dated the arrival of headlines about curved space. The "neglect of science" debate had helped to set the scene for a dispute, thanks to the rhetoric adopted by some of the more outspoken advocates of science.[62] Some colorful examples of this rhetoric may be found on the pages of *Science Progress*, a journal that its publisher, John Murray, wished to make more "popular."[63] Bowler identifies the market for *Science Progress* as "readers on the fringes of the scientific community, including those with a basic science training who were working in industry or as schoolteachers."[64] From 1912 until his death in 1932, the editorship was held by Ronald Ross, winner of a Nobel Prize in 1902 for his proof that malaria was transmitted by mosquitoes. Ross followed in the footsteps of Victorian reforming scientists like Joseph Norman Lockyer, the upstart solar physicist and founding editor of *Nature* who lobbied for increased state funding of British science. Ross also found time to paint and to write poetry and plays. For five shillings a quarter, subscribers to *Science Progress* were informed of the latest achievements in science and warned of the dire threats to civilization that had produced them. Ross had a clear sense of the two chief perils jeopardizing Britain's continued status as a great nation: lack of financial support for science, and a population in the thrall of "trumpery journalism, a contemptible stage, cinematograph shows, public-houses, silly processions, superstitions which call themselves religion, and streets full of untidy loafers with cigarettes in their mouths."[65] The scientists, poets, and philosophers who had helped to lift civilization out of savagery were left to die destitute, while "every kind of trickster or charlatan" was allowed to "enrich himself enormously at the public expense."[66]

During the early stages of the war, Ross crafted an allegory about "Mr. Man-in-the-Mass," an unsavory character who invites his pandering friends Field Marshal Militarius and Baron Politicus to survey the world, delighting in its liberation from the "fell magician" (evidently God).[67] "I am Emperor at last," exclaims the protagonist, "and my name is Alexander-Pompey-Caesar-Autocrat-Plutocrat-Democrat-Journalist-General-Admiral-Secretary-of-State-Hans-Jean-Bull-Smith-Jones-Robinson-the-Great, M.P." Swigging from his favorite bottle and waving the magician's wand, Mr. Man-in-the-Mass suddenly finds the world plunged into darkness and devastation. While the marshal and the baron tear at one another, "a tall old man with a long white beard" appears, dressed in "an old robe figured with quaint mathematical designs."[68] Chiding the "wicked Caliban" for having seized powers that he does not understand, the magician summons his sprites to chase the three troublemakers down from his mountain. For any readers who might have missed the subtleties of this allegory, Ross stressed that the technological and artistic wonders of the world had been created not by "men in the mass," but by "beings of another order who toil, plan, and think incessantly and who perish because of it."[69] In contrast, the masses did "nothing but sleep, eat, play silly games, and look at silly books containing silly pictures." Instead of a "passionate desire for truth," they were filled with superstition, base ideals, and "idle stuff which they read in worthless papers."[70]

How did Einstein's new theory of the universe fit into this Shakespearean vision of God as a mathematical magician, with scientists working as his sprites and the ignorant masses as Caliban?[71] Early notes and articles on the topic in *Science Progress* in 1915 and 1918–1919 were skeptical of Einstein's approach to space and time.[72] Aside from brief references to the verdicts of Lodge and Larmor (contributed by Harold Spencer Jones),[73] the journal fell silent on relativity during the periods when national press coverage was at its most frenzied. A short essay, "The Einstein Theory of Relativity," assured readers in April 1922 that Einstein had incorporated, rather than destroyed, Newtonian conceptions and credited Viscount Haldane's mathematical insight on the topic.[74] A subsequent note titled "Synthetic Relativity," contributed by J. R. Haldane (not the same person), outlined an alternative method of building a cosmological system that was said to be more in accord with traditional physics.[75] Einstein might have stirred up a good deal of interest, but he could not be entitled to Prospero's robes, for his symbols were not anchored in the created world. And yet, as the book review section of *Science Progress* made clear during the early 1920s, the "relativistic temple" was crowded with "a vast array of articles, papers, and books" on the subject.[76] These fell into two broad categories: the "popular or semi-popular type," and "treatises" that gave a "systematic and formal" account of the theory. Readers might be tempted by the "picturesque language and the semi-mystical, almost whimsical, flashes" of Arthur Eddington, or they could turn instead to the "business-like" style of those French writers who were able to "make straight the rough road to Relativity."[77] But for the "layman" any such "rigid exposition" remained out of reach: he would have to make do with loose understanding afforded by "simple analogies and familiar examples."[78] One review described a "very distinguished mathematician" who had not yet "grasped Einstein."[79] Assuring readers that this eminent figure "was, of course, quite capable," the reviewer explained that he was "so immersed in his own work that he had no time for the effort." The trouble was that "scientists of one group of science" were "unfortunately nearly always laymen as regards other groups." And yet the most interesting advances were being made by those who transgressed the boundaries between specialist areas: mathematicians who meddled in physics, for instance, or physical chemists. It was becoming harder for Prospero's sprites to keep up with one another.

The first place where readers with a scientific background would probably turn for help with relativity was *Nature*, published weekly at a price of 9d. (rising to 1s. as of January 1920). Prior to the eclipse results, the journal featured a friendly introduction to the new space and time: the text of Eddington's Friday Evening Discourse at the Royal Institution in February 1918.[80] This essay broached the Wonderland-style contraction and expansion of lengths when viewed by differently moving observers, and it prepared readers for curved space by suggesting that "you can almost see a living-picture of this real world reflected in a polished doorknob." Eddington concluded by mentioning the opportunity for eclipse experiments to test Einstein's theory the following year, and here the idea of light as an expensive commodity took a form different from that in its subsequent appearance under Davidson's name in the *Daily Express*. Calculating that a pound of light would cost something like £141,615,000, Eddington commented, "Fortunately, we get most of our light free of charge, and the sun showers down on the earth 160 tons daily. It is just as well we are not asked to pay for it." Where the subsequent *Express* moral of daylight saving would emphasize consumer thrift, Eddington's upmarket audience at the Royal Institution were reminded that God, unlike the electric light companies, made no charge for sunlight. This humorous note would have helped to alleviate furrowed brows following the introduction of Einstein's new law of gravitation, reassuring listeners that the traditional spirit of natural philosophy had not been completely abandoned.

What about Caliban? Ironically enough, the book that Arnold Bennett later dismissed as least likely to enable shared ownership of the new space and time was serialized in a radical socialist newspaper prior to its publication in book form. Extracts from Bertrand Russell's *ABC of Relativity* were published during March 1925 in the *New Leader*, the 2d. weekly paper of the Independent Labour Party. A striking affinity between the language of British socialism and the new space and time emerges in this context, giving Einstein's theory significance as part of a broader movement to eradicate inherited privilege and transcend private interests. Coincidences of terminology between the discussion of economic questions and Russell's exposition would have made for a reading experience very different from that of readers who purchased Russell's *ABC* at 4s. 6d. or borrowed it from a library.[81] Detached from their place in his witty yet challenging book-length exposition, Russell's wriggling measuring rods and hills in space-time would have been even harder for *New Leader* readers to make sense of. But the Independent Labour Party context helped to make the grasping of theoretical detail less urgent than an appreciation of social and economic progress. The resulting popularization is diametrically opposed to *Conquest*'s sovereign aesthetic. On the pages of the *New Leader*, relativity was allowed to resonate with the language of economic and political reform, without being imposed as a model or pattern for the management of human affairs.

The Parochial Prejudice of Earth-Dwellers

Ramsay Macdonald became Britain's first Labour prime minister in January 1924. He held office for less than a year, but returned from 1929 to 1935. An industry of popular relativity books sprang up during this period, and it was impossible to think, read, or write about a new theory of the universe without wondering about how it might be used to support—or counter—alternative political systems. By the end of the 1920s, exposition of the new physics had acquired spiritual and idealist associations, and refutations of scientific materialism were associated with bourgeois alarm at the prospect of a Marxist uprising. Reviewers and critics during the 1930s complained about the unscientific burdens that the new physics was being asked to carry, and they worried about readers who might not be able to tell where science ended and ideology began.[82] A reconstruction of the reading experience made available through the serialization of excerpts from Russell's *ABC of Relativity* reveals a less fretful construction of the mathematically innocent reader's capabilities.

The *New Leader* was edited from 1922 to 1926 by Henry Noel Brailsford, a committed internationalist and former professor of logic whose radical socialism conflicted with Ramsay Macdonald's vision for Labour in Britain.[83] Macdonald also resented Brailsford's "highbrow" aspirations for the Independent Labour Party organ, which featured literary, historical, and philosophical contributions aimed at "raising the whole intellectual standard of the Labour movement."[84] During 1925 Brailsford produced a series of articles restating the case for socialist transformation, with headings such as "Who Shall Ration Work?," "The Strategical Roads to Power," "Socialism and Property," and "Industry as a Public Service."[85] Another prominent story at this time was the Bank of England's return to the gold standard. The equivalence between a pound note and a gold sovereign had been abandoned in 1914 to help finance war, and postwar rises in the bank rate were helping to push up the gold value of paper money so that Britain's war debt to America could be paid off more quickly. The *New Leader*'s regular satirist "Yaffle" commented on "the mass of the people" being left "in total ignorance" regarding this process.[86] "People read a lot about the Gold Standard in the papers," he noted, "but they only learn a few of its characteristics; they do not learn what it is." Most had grasped "that it is something one returns to, or does not, or that is, or is not, thoroughly undesirable." But, as Yaffle observed, "such attributes are equally applicable to Margate or the Low Waist."

New Leader subscribers could set themselves apart from the mass of newspaper readers thanks to a clear exposition of the gold standard by Barbara Wootton, which appeared the same week as part 2 of Russell's *ABC of Relativity*.[87] The paper's cover for that issue featured St. George preparing to "look that dollar in the face again," mounting his steed on the face of a twenty-shilling coin. But Labour, represented by a man in a cloth cap, pointed to a slavering black creature with "Unemployment" on its wings: "Hold hard. You haven't slain your dragon yet." Inside, beneath the heading "Our Autocratic Bankers—Human Sacrifices to the Golden Calf," Wootton explored the implications of a decision that had been "reached in the quiet of the Bank's buildings by two dozen men who, whatever their sense of public duty, are in no way responsible to the community." There were immediate grounds for comparison between esoteric financial decision making and the obscure mathematics of relativity, a connection that hardly needed spelling out for *New Leader* readers. But a striking blend of congruence and contradiction between Russell's exposition and the paper's coverage of economic questions offered readers a broader perspective on cosmic abstraction.

Headed "Why Clocks and Footrules Mislead," Russell's first article opened with a clarification of what relativity meant.[88] "People often imagine that the new theory proves *everything* to be relative," he noted, "whereas, on the contrary, it is wholly concerned to exclude what is relative and arrive at a statement of physical laws that shall in no way depend upon the circumstances of the observer." The extent to which circumstance affected measurement had turned out to be much greater than physicists had previously believed, and Einstein's great achievement lay in showing "how to discount this effect completely." The second article, "How Space and Time Are One," explained the relativity of time.[89] "Different clocks give different estimates of the time, according to the way they have been moving," said Russell, assuring readers that "there is no standard by which we can say that one of these clocks is right and the others are wrong." A subheading, "The Parochial Prejudice of Earth-Dwellers," introduced the problem of deciding whether two events, one on the earth and another on the sun, happened at the same time. "If you were on a comet moving very rapidly relatively to the earth and the sun," Russell

3. *New Leader*, March 13, 1925. Reproduced by kind permission of the Independent Labour Party (www.independentlabour.org.uk) and the Working Class Movement Library, Manchester.

explained, "you would form a different estimate (after allowing for the speed of light) than you would from an observatory on the earth." To say that we are right and that a being on a comet is wrong in their estimate of a time interval "is merely the parochial prejudice of people living on the earth," he concluded. "There is no right or wrong about it." Einstein had shown that measurements of distance and time had no objective reality, but Russell affirmed that there was a measure, called the "interval," that was the same for all observers and could therefore be "taken as a reality of physical science."

In part 3, "The Eel and the Measuring Rod," Russell moved from Einstein's 1905 special theory of relativity to his 1915 treatment of gravitation.[90] This difficult development was broached through the idea that geometry had now become mixed up with physics, resulting in the abandonment of straight lines:

> We are apt to think that, for careful measurements, it is better to use a steel rod than a live eel. This is a mistake. To an observer in a suitable state of motion the eel would appear as rigid as the rod does to us, while the rod would appear to be constantly wriggling. Nor can it be said that such an observer would see things falsely, while we see them truly. . . . Measurements of distances and times do not reveal properties of the things measured, but relations of the things to the measurer. What observation can tell us about the physical world is therefore more abstract that we have hitherto believed.

Here, as in the discussion of time, the abstraction of relativity from common experience is tied up with a theme that ran through coverage of economic and social questions in the *New Leader*: eliminating prejudice.

Under the provocative heading "Nature, the Anarchist," Russell's final article explored the implications of relativity for the conduct of our lives.[91] Dismissing as "merely slavish" the search for moral guidance in the laws of nature, he noted that "if nature is to be our model, it would seem that the anarchists will have the best of the argument." A section subheaded "The End of Monarchy" compared the Newtonian sun to "a monarch whose behests the planets have to obey," whereas in Einstein's theory "the planets never notice the sun, but adopt the easiest course at each moment, like water running down a hill." Russell adapted this point from Eddington's claim that "the longest time must be taken over any job," shearing away the astronomer's joke about "the great trade union of matter." In Russell's version, the resulting message was a "law of cosmic laziness" in a world characterized by individualism: "The physical universe is orderly," Russell observed, "not because there is a central government, but because every body minds its own business." Reaffirming the "extremely abstract" quality of this new theory, Russell conjectured that "we are probably not at the end of the process of stripping away what is merely imagination, in order to reach the core of true scientific knowledge." Abstraction might be difficult, he conceded, but it is "the source of power." A financial analogy was offered, aimed at reconciling readers to this pragmatic view: "A financier can deal in wheat or cotton without needing ever to have seen either; all he needs to know is whether they will go up or down." Similarly, the physicist "knows nothing of matter except certain laws of its movements, but this knowledge is enough to enable him to manipulate matter." It was, Russell concluded, "astonishing that so little knowledge can give us so much power."

With its emphasis on individualism, the metaphor from capitalist finance, and an explicit rejection of "slavish" models for human conduct based on natural law, Russell's *ABC* clearly eschewed any reduction of relativity to socialism or vice versa. But his arguments against parochialism and his questioning of standards for measurement would have appealed to *New Leader* readers who shared the paper's commitment to economic reform along socialist lines. In the same issue as Russell's first article, "Why Clocks and Footrules Mislead," Lewis Donaldson (the canon of Peterborough) criticized the deans of Durham and St. Paul's for their recent attacks on the railwaymen and trade unions.[92] Noting that prejudices were "gathered up unconsciously in childhood and onwards from our traditions, surroundings, customary colleagues and acquaintance," Donaldson called for scholars and divines to "shed that subtle class bias which obscures from our perception the awful injustice of our present social order." Brailsford's essay "Socialism and Property" appeared on the next page, arguing that "the glaring inequalities of to-day" could not simply be replaced with "a rigid equality of incomes." The fundamental principle of the living wage required that every worker be paid according to their need, for a socialist society could not allow "the same wage or salary to be paid to the childless worker and to the father of a family." There could be no other way to determine wages: "By what measuring rod will you decide the relative value of the social services of a mother, a poet, and a carpenter?" While this rhetoric of variable measurement chimed with Russell's wriggling measuring rod, Barbara Wootton's emphasis on fixed standards worked in the opposite direction. She argued that money should "always have the same value, just as a yard, which is a measure of length, always has the same length." A return to the gold standard was unlikely to establish this kind of stability, for the value of gold had been variable before the war and was likely to remain unstable unless "wisely managed by international action."

The disjunction between Russell's wriggling measuring rod and Wootton's insistence on a yard always having the same length supports the philosopher's rejection of slavish correlation between natural law and human affairs. Beyond this superficial difference, however, Russell's and Wootton's articles share a broader message: that unstable measures must everywhere be detected and rejected in search of more reliable standards. The *New Leader* context would have strongly encouraged this verdict, leaving readers to conclude that Einstein was simply participating in an age of steadily receding parochialism. Any anxiety on the reader's part about not understanding relativity would be eased by locating it within a broader campaign to eliminate prejudice in all its forms, from the unconscious minds of deans to the decisions made behind closed doors at the Bank of England.

An earlier, more explicit connection between the gold standard and Einstein's theory treated its readers rather differently. *Credit-Power and Democracy*, by Major C. H. Douglas, was serialized during 1920 in the *New Age*, a weekly paper costing 7d. and aimed at intellectuals who wanted to be kept abreast of the latest developments in politics, art, and literature. Here Douglas expounded Social Credit, a system for the equal distribution of economic power as a means of preventing future war. In his twelfth chapter he dismissed "the popular delusion that a gold or other standard is an absolute measure of value," but unlike Wootton, he argued that such absolutes were neither possible nor needful.[93] Appealing to "readers who are familiar with the mathematical hypothesis known as the theory of relativity," he reminded them that "it is impossible by means of physical measurements to determine the absolute velocity of a body through space." Douglas refrained from spelling out the connection to his economic argument, noting simply that "certain analogies will no doubt present themselves." Turning his attention to "the average person not particularly interested in such matters," he presumed that they would experience "no difficulty . . . in grasping what is meant by 'ten miles per hour,'" even though they "cannot conceive of 'a mile' as distinct from 'a mile long.'" Here the explicit comparison with Einstein's relativity works to assure *New Age* readers that Douglas's economic arguments are complex and progressive, while the appeal to an intuitive grasp of speed implies that Social Credit has not lost touch with common experience. The majority of *New Age* readers might feel more at home with "ten miles per hour" than they did with "the absolute velocity of a body through space," but that would not prevent them from assuming the vantage point offered by Douglas.

The *New Leader* was exceptional in carrying no implicit or explicit division between the capacities of experts and their wider audience. It was able to achieve this because authors and readers shared a commitment to socialist principles that shaped their outlook on any topic. A contrasting example of alignment between expertise and innocence emerged at the end of the 1920s in a new publication occupying the opposite end of the political spectrum: a popular science magazine with an aesthetic shaped by consumer individualism.

There's Relativity in Everything

Every month from April 1929 onward, the bright yellow cover design of *Armchair Science* carried the same simple image: a handsome young man peering into his microscope, connected by chain links to a similarly smart fellow perusing the magazine from the comfort of his armchair. This same motif could be discerned on the front cover of his own copy, suggesting worlds within worlds of debonair scientific practitioners dangling from the armchairs of equally fashionable readers. The effect, sometimes referred to as *mise-en-abyme*, would have been familiar to readers from a well-known 1920s advertisement for Quaker Oats, besides many other commodities from cocoa to tobacco.[94] Its use on the cover of a science magazine signaled the editor's awareness of his readers as consumers well used to marketing strategies and capable of making their own choices, in science as much as in their domestic or professional lives. For 7d. a month (rising to a shilling in 1931 but then going down to sixpence later in the 1930s), readers could possess their own personal worlds of science from the comfort of their easy chairs, without the need for taxing visits to laboratories or workshops. Article titles emphasized the human interest of each topic: "The Romance of Rubber," "The Future of Women," "The Secrets of a Broadcasting Studio," "The Mystery of Sleep," and—of course—"Learn to Fly from an Armchair."[95] The team behind *Armchair Science* exuded a blend of authority and plain talk even in their names: director Lt.-Col. J. T. C. Moore-Brabazon M.C., technical advisor Professor A. M. Low, and editor A. Percy Bradley (M.I.A.E., A.M.I., Mech. E.). Bowler notes that "Low was despised by the scientific community, partly because he was not really a professor but also because he represented exactly the kind of individualistic, tinkering practical expertise which the professionals hoped to transcend in their bid for influence with government and industry."[96]

Readers of *Armchair Science* were actively encouraged to air their own opinions, not only through the speculative, personal style of the articles but also by direct appeals from authors and by invitations to enter competitions. In February 1930 the handsome prize of two guineas was offered for the best letter giving reasons for or against belief in psychic phenomena, since "no definite decision has ever been reached" by sci-

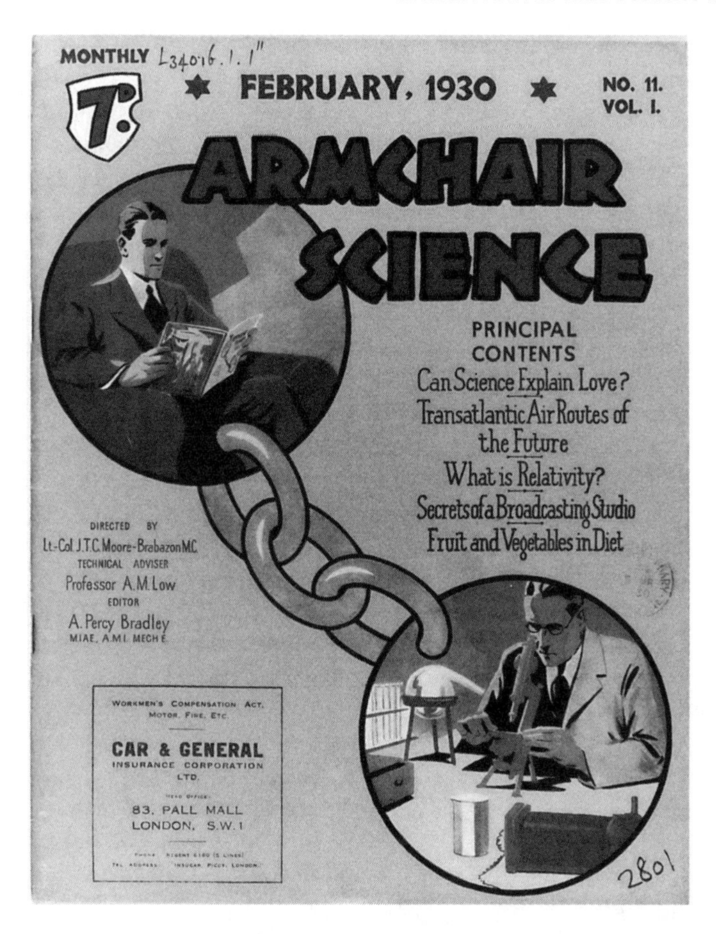

4. *Armchair Science*, February 1930. Reproduced by kind permission of the Syndics of Cambridge University Library, L340:1.b.1.

entific authorities.[97] In "Physics and the Ordinary Man" in August 1930, Moore-Brabazon addressed the "many readers like myself who do not profess to be mathematicians but who like to follow current thought in physics," declaring that he would "like to know from them what is their attitude and whether they do in any way share my own views."[98] In "Some Thoughts on Current Physics," published in February 1931, he reported that one recent popular physics book had "sold better than any novel."[99] This was, he declared "great news," for it justified "the ordinary man in the street (as I claim to be) in reviewing in a few sentences the position to which we have got at the present time." Turning to "some of the difficulties which must strike the man in the street," Moore-Brabazon addressed the problem of trying to "visualise the world as a four-dimensional continuum."[100] In a finite universe, any journey through space must eventually return to its point of origin. Surely a journey in time as the fourth dimension ought to have the same outcome? Moore-Brabazon reasoned that "the infinite future and the infinite past" must meet, and yet he had "not seen this conception discussed by anybody." Illustrations reinforced the claim that *Armchair Science* had "forestalled, in fact, the fashion which is now being exploited by great scientific leaders of writing articles in a popular form."[101] The top left corner of the first page presented a photographic portrait of a surprisingly youthful Moore-Brabazon, in diagonal opposition to a rather more staid-looking Sir James Jeans. On the facing page was an image of Sir Arthur Eddington, looking remarkably like Moore-Brabazon himself. The fourth-dimensional suggestion for joining up the past and future was apparently corroborated later on that year, when the magazine carried "Can We Travel in the Time Dimension?" by J. W. Dunne, summarizing the author's alternative theory of time (published in his 1927 book *An Experiment with Time*).[102] Identifying themselves unequivocally with the man in the street, *Armchair Science* contributors presented scientific authorities as beholden to the imaginative capacities of ordinary readers—dangling by a chain from armchairs across Britain.

An unsigned article headed "Relativity" in February 1930 assured readers that "lack of advanced knowledge of mathematics forms no bar to the understanding of the simpler principles of relativity."[103] The author, probably Low, aimed to show how "the idea of relativity enters in a practical way into our daily lives" and made free use of motorcars, greyhound races, and bombs to bring Einstein within reach. The spirit and content of this article were very similar to what was seen in *Tit-Bits* a decade earlier, and the result was an entertaining pastiche of Einstein exposition.[104] The author used what were by now familiar devices (vehicles, planets, weapons and clocks) to turn away from four-dimensional, curved space-time, shifting the focus toward rediscovery of a common-sense outlook. Invited to picture "two greyhounds running round Wembley in opposite directions," readers were presented with two competing interpretations. "To an observer stationed in the centre of the ring, neither of the dogs move closer or further away. In this respect they may be considered stationary." But, "by taking the sun as a 'fixed' standard," so that "the dogs, Wembley and the earth will all be travelling round the sun," it became apparent that the two dogs "are really both going about nineteen miles a second *in the same* direction." The most significant perspective, however, was found to be "our judgement of the relative speeds of dogs travelling in the same direction round the track," which was, the author acknowledged, often "a matter of considerable commercial gain or loss." The article concluded with a more dramatic version of L. G. Brazier's verdict in *Conquest*.[105] Unlike electricity and wireless, Einstein's theory was not bringing about any "immediate and obvious changes," but in broader terms it had contributed to an "age of Science" in which mankind was increasingly alive to nature's indifference. The realization that "our earth is a tiny planet in infinite space, through which roll mighty Spheres in unending cycles" showed that human creatures were "small fry in this gigantic scheme." Relativity was incorporated within a staple theme of popular astronomy, the vast emptiness of space: "Nature goes on her way, regardless of our lives with their hopes, fears and beliefs." The "unending cycles" of *Armchair Science* were more in keeping with classical depictions of the universe as a machine than with Einstein's curved space-time.

In contrast to expository books on relativity, whose authors attempted to renegotiate the relationship between common experience and mathematical abstraction, articles in *Armchair Science* tended to eliminate the need for negotiation in the first place by according equivalent status to scientific theory and individual imagination. This was largely achieved by tactics very similar to those found in the *Daily Express*: the focusing of any given topic through an appeal to the senses. A February 1932 article by Professor Low, "There's Relativity in Everything," featured photographs of racing cars, a lighthouse, war graves, and a giant three-hundred-year-old tortoise carrying three baby giant tortoises on its back. Low began by proclaiming that "no one can possess definite knowledge," insisting that this was the "idea of relativity" that should remain "foremost in our minds when examining any scientific problem whatsoever, for facts do not exist other than as a function of time."[106] Observing that one person's thousand years might seem a mere fraction of a second to another, he reminded readers of the basic principle that every experience could be viewed from several different perspectives. "When I see some particularly brutal murder," he mused cheerfully, "I often say to myself that it may be necessary to kill a few apparently innocent people in order that from their ever-living thoughts, and their rotting bodies which fertilise the soil, more life may be created." It was important, he urged, to remember that "we live like butterflies, and that our viewpoint depends upon the speed at which we view the world." The cinematic associations of distorted space and time provided a further opportunity for putting life in perspective. Observing that "light and heat take time to reach us from the sun," Low proposed that "the path of that light is not always straight, but rather is drawn out just like a vast cinematograph picture would look if we were falling from the roof to the floor as the film was shown."[107] He then invited readers to imagine they were watching a series of photographs taken of the earth at intervals of several years: "like a film of which the man in the sun was the producer, we should see buildings falling into heaps of dust, whole forests torn to pieces by invisible insects, and we should say: What will happen next!"

Low's writing frequently defies conventional logic, proceeding by leaps of association and never missing an opportunity to relish unpleasant or extreme physical experiences. The result is a parody of science writing, complete with broader claims about the nature and status of scientific knowledge. Discussing the proliferation of "bio-chemistry," "physics of chemistry," and the "electricity department of chemistry," he marveled that "we have even coined beautiful words like electro-chemical-physics to show that if we do not know any facts, we can at least make an excellent pretence." In *Science Progress*, increasing specialization was shown to be making new science inaccessible even to other scientists. But in *Armchair Science* the proliferation of hyphenated subdisciplines was license for anybody to invent their own department of research. Low even expounded a theory of love waves as a hitherto undetected part of the electromagnetic spectrum in February 1930.[108] This climate of invention meant that anything was possible. "Let no one . . . think that a departure from religion towards physics is going to take him into a realm less mystical than heretofore," Moore-Brabazon warned in "Physics and the Ordinary Man," conjecturing that no previous period in physics had been "of more interest than the present, for everything is now in the melting-pot."[109]

Popular physics writing during the late 1920s and early 1930s gave Caliban access to Prospero's magic, but as J. J. Thomson regretted in his memoir, there were now as many wands in circulation as there were individuals. In Thomson's view, the sheer variety of universes available by 1930 was grounds for mistrusting every one of them: "We have Einstein's space, de Sitter's space, expanding universes, contracting universes, vibrating universes, mysterious universes. In fact the pure mathematician may create universes just by writing down an equation, and indeed if he is an individualist he can have a universe of his own."[110] Thomson's protest extends beyond a distaste for mathematical speculation to include a publishing phenomenon that had come to play a defining role in public perceptions of science in Britain. If each reader was licensed to create his or her own universe, where did that leave scientific authority?

THREE

Cracks in the Cosmos: Space and Time in Pulp Fiction

Along with sensational news headlines, the main problem facing Percy Harris and his writers was "uninstructed gossip."[1] However many pages of illustration and explanation a magazine offered its readers, expositors had no control over what happened to science themes when they were drawn into informal conversation. An ideal resource for studying Einstein's relativity in British popular culture would be the raw contents of Harris's postbag for the early 1920s. Without access to the queries and demands that poured onto his desk, the closest we can get to mathematically innocent engagement with the new cosmology is by searching ephemeral fiction magazines of the period for depictions of characters reading and talking about space and time. These magazines do not necessarily yield a realistic depiction of how relativity impinged on everyday experience—with their lethal rays, cracks in the cosmos, and dematerializing men, these Einstein-themed adventure stories would have appealed precisely for their departure from humdrum existence. But with their eye for contemporary satirical detail and their acute awareness of fiction readers' own specialist skills as story consumers, authors of British pulp fiction were able to provide access to Einstein's universe on terms that were not available to science popularizers.

The richest source of fourth-dimensional adventure stories during the early 1920s is the *Red Magazine*, published fortnightly by Amalgamated Press.[2] Despite the socialist connotations of its name, the *Red* was a mainstream magazine owned by the Harmsworth family.[3] A striking trend emerges from the spate of Einstein-themed tales that appeared in the *Red* between 1920 and 1923. Relativity was not initially accommodated within science fiction plots, despite clear precedents within that genre. Time travel and journeys at the speed of light were already familiar to the reading public, with pulp authors prior to 1919 taking inspiration from H. G. Wells and Camille Flammarion.[4] Einstein themes were picked up by authors who were simultaneously producing science fiction, but the new cosmology was incorporated within supernatural plots instead. These uncanny adventure stories depict mathematically innocent characters having direct physical experience of Einstein's universe, of the kind outlawed in magazine exposition. But they invariably conclude with a warning that the unsettling new cosmos is best left uninvestigated, dismissing its features as a figment of dreams or some other mental aberration. Why did intimate contact with the fourth dimension, craved by the importunate readers of *Conquest*, turn out on closer inspection to be so hazardous?

The Midst of Utter Barbarism

Relativity themes were ideally suited to the comic, otherworldly adventure writings of Coutts Brisbane, an Australian author based in London.[5] Equally at home with naval exploits, animal fables, crime, Arabian romance, and the jelly inhabitants of Mercury, Brisbane was a prolific contributor to the *Red* (launched in 1908) and its companion magazine, the *Yellow* (launched in 1921). These magazines were published on alternate Fridays at a price of 7d., so fans never had long to wait for a fresh escapade. A skilled writer, Brisbane enlivened conventional plots by introducing contemporary satirical twists and hybridizing subgenres. "Mr. Fipkins and the Younger World," published in October 1923, draws together the key strategies that were applied to relativity during the first few years of its life in popular fiction: sartorial interests, adventure over theory, and denial of evidence. The story is set in offices just off Fleet Street, where Mr. Fipkins and the narrator, a man named Smyth, both run print supply agencies. Smyth confesses to having "fought shy" of his business neighbor, whose philosophical interests make him a somewhat tedious lunchtime companion.[6] Fipkins's reading includes "Kant and Hegel, Spencer and Nietzsche," and he has "even wrestled with and taken a fall from Einstein." He sports a "top-hat and morning-coat of an excellent style," old-fashioned items of dress that were "once the common wear of the City" but are now "the exception rather than the rule." This relic of a former business age is, we learn, also a man of *embonpoint* who has not glimpsed his own feet for some years.

Gazing out of the window one morning, Smyth is astonished to see some bricks turn transparent, revealing sunlit landscape, grassland, and thick forest beyond. Along comes Fipkins, arriving on the dot of 9.35 as ever, his "top-hat shining, morning-coat a thing of sartorial beauty, his pearl-grey spats carefully fitted about carefully polished boots."[7] And he vanishes. As Smyth investigates the now completely solid brick wall, Fipkins suddenly reappears, his former "spruce and debonair" attire "gone miraculously to seed." Every marker of civilization has been torn to shreds: his hat is "battered and shapeless . . . with part of the brim gone and all of the gloss"; the "pearl-grey spats," half one trouser leg, and his umbrella are missing; his shirt and the remains of his coat are filthy. Fipkins has also acquired "a short club or hatchet, with a stone head bound to the haft with thongs," and his shoulders are draped in a "fur hearthrug." Smyth hurries his "marvellously dishevelled" associate inside, where Fipkins paws the arms of his chair in a wild manner and insists that Einstein "and all those other fellows who talk about the Fourth Dimension" are right.[8] He has fallen through a gap in the cosmos, from "the heart of civilisation" to "the midst of utter barbarism," spending three hours at large in a primitive world while barely three minutes have passed in present-day Fleet Street. Thanks to a box of matches discovered in his pocket and some judicious brandishing of his umbrella, Fipkins has managed to escape from a pack of hungry wolves, evade a ferocious saber-toothed tiger, and outwit a horde of Stone Age villagers.

Having initially supposed the strange transparency of the wall to be a "trick advertisement," Fipkins is now convinced that he actually stepped onto another plane where time was passing at a different rate and London was four thousand years behind the present.[9] It is not just Fipkins's attire that suffered during his three hours in primitive London: the survival instinct quickly reduced him to theft and violence in the younger world. Imprisoned by a jealous village medicine man, the Fleet Street agent appropriated a holy animal fur, club, and necklace of bones, hoping that these trappings would make the common people fear and respect him. He then cracked a few skulls and thrust a spear into any villagers who attempted to impede his escape. It is only once he has regained the safety of Smyth's office that Fipkins begins to wonder whether "some sort of truce" might not have been arranged.[10] Anxious at the prospect of a time-traveling tiger, Smyth is comforted by the knowledge that there is a gunsmith nearby. His mood lightens at the thought of an encounter between the "eminently staid and respectable constables" of Fleet Street and "a squad of wild men from the Stone Age."[11] But the crack in the cosmos has now closed.

Apart from the protagonist's enthusiasm for Einstein, is there anything distinctively "relativistic" about this story? Could it have been written before 1919? Two features distinguish Fipkins's adventure from a Wellsian time-traveling narrative. The first is that Brisbane's character has no device for navigating the dimension of time. He is plunged into the younger world by accident and is almost lost from Fleet Street forever. His "last glimpse of the other world" reveals "everything tilted, as though in the time I had been there that plane and this had ceased to coincide," and he concludes that a longer absence might have meant slipping through the aperture "only to find myself nowhere."[12] Stories of accidental access to parallel worlds were not uncommon before 1919, but the prospect of ending up "nowhere" gives this adventure a distinctive contemporary edge.[13] (While it is true that the schoolmaster in H. G. Wells's "Plattner Story" ends up in the Other World by accident, that has more to do with his "temerarious" approach to the mysterious green powder than with any innate property of the cosmos.) The second distinguishing feature is a continual emphasis on the proximity of the two vastly separated ages. Fipkins's excellent knowledge of London geography enables him to keep his bearings in the "younger world," climbing a tree to evade wolves in Fetter Lane and sending the tiger "full speed in the direction of Farringdon Street Station."[14] The narrator's anxiety about tigers and his amusement at a possible confrontation between bobbies and savages add to the sense of primitive and civilized worlds bleeding into each other.

This sense of accidental transition translates the difficulty of mathematical theory into unpredictable hazards over which the characters have no control. In a science fiction narrative, readers would feel confident that an explanation exists, even if it were not actually given. But "Mr. Fipkins and the Younger World" asserts mystery and action over the possibility of exposition. When Fipkins launches into an account of how an inch-long bullet might really be six inches long when traveling at fifteen hundred feet per second, Smyth is convinced that "there must be a fallacy somewhere."[15] He is far more anxious to hear more about "wolves and other appurtenances of the affair," and asks for technicalities to be set aside. At the story's end he confesses that "my head begins to reel when I try to conceive of fourth or fifth dimension planes lying alongside our world" and declares Fipkins "wiser" for having abandoned his convoluted theories and resigned himself to treating the incident as a dream.[16] This is the standard ending for Einstein-themed stories in the *Red*.

The mixing up of space and time associated with relativity provided an opportunity for storytellers like Brisbane to adapt the Wellsian time-travel plot, shifting its emphasis from evolutionary tourism to postwar satire on the savagery underlying modern life.[17] The war had produced a good deal of commentary on this topic, and various solutions had been proposed for understanding and remedying the tragedy of modern barbarism.[18] The adventure of Mr. Fipkins uses Einstein themes to pastiche the idea that inside every modern man there lurks a violent savage. Mike Ashley has noted that although the *Red* was "designed as an adult all-fiction magazine," it "ran many stories by boys' adventure writers and often read only a little above a boy's story-paper."[19] This may even, he observes, "have been the secret of its success": appealing to adult readers who found respite in its generous provision of "escapist fun and romance."[20] The magazine's affiliation with fiction for boys enabled contributors to take a light-hearted approach, producing stories in which savage behavior could remain entertaining despite the recent carnage in Europe. Here an author like Coutts Brisbane could portray the idea that every city worker secretly longed to brandish spears, tear his clothes, and escape from everyday routine without having to condemn this as a dire threat to civilization. But the risk of being stuck "nowhere" enables the story to voice other doubts about male identity—for example, about the role of a well-dressed and portly shop manager in postwar British society. The satire on dress and conduct allows the idea of the inner savage to remain a compelling fantasy while at the same time suggesting that modern theories about primitive regression cannot answer the question of how to be a man after the Great War. This was a question that, like "What is Einstein's theory?," had no easy or immediate answer.

A more sedate version of the younger London story appeared in *Mystery-Story Magazine* for April 1924, penned by Captain Lancelot de Giberne Sieveking (godson of G. K. Chesterton).[21] Describing its author's style as "strongly reminiscent" of H. G. Wells, the editor introduced "The Lost Omnibus" as "an entertaining travesty of the theory of the fourth dimension."[22] A London bus following a diversion is accidentally shifted in time, but in place of Wells's monster crabs on a degenerate future shore, or Brisbane's saber-toothed tigers bound for Farringdon Station, passengers glimpse Queen Victoria in a much cleaner, brighter Oxford Street, where hansom cabs are drawn by horses in sparkling harnesses. A passing gentleman's dress does not appear so very different, but the girl with him looks peculiar in "mutton-chop sleeves" and a "great skirt flowing out behind her, dragging in the dirt."[23] With their obsessive attention to costume and convention, pulp authors acknowledged the ability of readers to locate the present moment on their own terms, without the need for closely printed columns of exposition in the *Times* or headache-inducing treatises on space and time. Disrupting faith in one straight line from the past to the future, Einstein's theory affirmed what authors and readers of fiction already knew: that the present moment was, confusingly, going forward in some respects and backward in others. Oxford Street might be dirtier, but women's clothing had become more practical. As in the newspaper coverage, this was experienced as a disruptive relativization of Darwinian evolution. But where the *Times* drew a moral about voters, the pulp authors acknowledged a rather different form of expertise on the part of their readership.

One Peep at the Future

Stories in pulp magazines display an arch awareness of their readers as consumers of fashion, knowledge, and technology. As in the popular science magazines, advertising carried alongside the stories helps establish an overall reading experience, constructing a "subject position" for the reader in relation to the magazine's contributors and their creations.[24] But there is a higher degree of complicity evident in the *Red* and the *Yellow* than in *Conquest*, illustrated by an example involving curative electricity devices. Advertisements for these products were often aimed at men or women specifically. "5,000 WOMEN CURED IN RECORD TIME," announced the Electrological Institute (located at Vulcan House on Ludgate Hill).[25] The copy declared that women suffering from a variety of ailments, including neuralgia and constipation, "will hail with joy the announcement of scientists who *prove* that in the use of Curative Electricity lies the solution to all their Health Troubles." The illustration showed a fashionably bobbed girl perched on a suburban garden gate with a determined look in her eyes, hockey stick and tennis racket at her feet. Lightning bolts encircle her head and dance on the palm of her upraised hand. An equivalent male advertisement announced, "Neurasthenia (Nerve Weakness) Cured by Electricity," listing a wide range of symptoms: indecision and lack of self-confidence, dread of open or closed spaces, blushing, turning pale, shrinking from strange company, sudden impulses, and a craving for stimulants or drugs.[26] The illustration depicted an old-looking man flinching at the approach of a motor vehicle, in contrast to a much sturdier, gleaming figure surrounded by darts of electrical power.

In a playful editorial touch, the *Red* positioned a story by A. E. Ashford, which refers to one of these devices, immediately before a British Electric Institute advertisement for the Ajax Dry Body Battery Cell, designed to restore "proper manly strength."[27] Inaugurating a run of temporal adventures in the *Red* during 1923, "The Time-Adjuster" features a stockbroker named Henry Carslake, who acquires a curious watch with "little raised bosses" on the wristband. It is said to resemble "the small battery affairs on the inside of the electric body belts which used to be so much advertised at one time," a near-futuristic reference that might cheer any reader whose own Ajax lay neglected in a drawer.[28] The watch, however, has an opposite effect on its wearer's manly presence. Repeatedly described as a "bulky figure," Carslake is said to have an "eminently reassuring" and "Everydayish" personal presence—he is a "big, bluff, breezy fellow" with no trace of "spookiness." And yet a single flick of the device causes him to dematerialize, leaving only a hat of "grey-green velour" and a pair of gloves behind in the narrator's office.[29] This story parodies *The Time Machine* by placing the time-travel device in the hands of a London stockbroker. Where a philosopher of broad vision might choose to witness battles and coronations, or voyage from the dawn to the demise of humanity, Carslake uses the Time-Adjuster (its name suggests that the fourth dimension may soon be absorbed by the insurance industry) to intercept an urgent business letter, recover a lost recipe for a miracle cold cure, and purchase shares in a South African mining company just before they boom. Set in London, this story was originally written for the American fiction market, and it differs from the other British stories by having its protagonists focus on efficiency, advertising, and investment.[30] Materializing hatless in the street, Carslake causes double distress to a junior clerk who has been charged with delivering a letter to his home address. By traveling in time to intercept the letter, he has, in effect, robbed Dolliver's errand of its purpose. The clerk is "overwhelmed with remorse at the thought of the useful work he might have accomplished" instead of wasting time and taxi fares in a "deplorable lapse" of efficiency.[31] This teasing of what were held to be typically American values forms a contrast to Smyth and Fipkins, who are more concerned with luncheon and restorative glasses of whisky.

Fipkins and Carslake travel into the deep and recent past, respectively, but Brisbane had explored futuristic encounters in the summer of 1920. His story "Thus Said Pel!" envisages London under the despotic rule of

5,000 WOMEN CURED IN RECORD TIME

Remarkable Success of Electrical Treatment of Women's Ailments.

WOMEN who suffer from Nerves, from Anæmia, from Functional Disorders, Neuralgia, disorders of the Digestion, Constipation, Liver and Kidney troubles, will hail with joy the announcement of Scientists who *prove* that in the use of Curative Electricity lies the solution to all their Health Troubles.

Think of the joyousness of being transported from Ill-health, unnecessary tears, depression, bodily pain and tremulous "nerves," to vigorous Health! Think of the miracle of feeling Well and Happy after discomfort and despondency! Think of the possession of clear complexion, bright eyes and springy step after those interludes when you really "wished you were dead."

Such is the power of modern Curative Electricity, as introduced to thousands of Physicians, their Patients, Hospitals and Private People. Truly the discovery has come just in time. Anyone can adopt it. The results are *proved* beyond dispute. There are no drugs, patent medicines, pills or nauseating formulas—simply the Life-Saving, Nerve-Strengthening, Blood-Creating action of Curative Medical Electricity by means of special appliances which any lady reader can use with the most trifling expense and not the slightest interference with the daily round.

Lady readers should make a special point of calling to see the Chief Consultant at The Electrological Institute or of writing for a copy of the FREE BOOK on Women's Ailments. There is also a booklet for gentlemen. Write to-day to **The Superintendent, The Electrological Institute, 175, Vulcan House, 56, Ludgate Hill, London, E.C.4.** (J. L. Pulvermacher & Co., Ltd.)

5. Female curative electricity advertisement, 1925. *Yellow Magazine*, vol. 16, no. 101 (July 24, 1925): 331. The Bodleian Libraries, University of Oxford, Per.25612.e.534.

NEURASTHENIA

(NERVE WEAKNESS)

CURED BY ELECTRICITY.

To-day the conditions of life are causing a serious increase in Neurasthenia and other Nervous and Functional Disorders.

The symptoms of Neurasthenia are many and varied. They are mainly mental or nervous, and often the victim is quite unaware of the fact that he or she is travelling rapidly towards Nervous Exhaustion and Nervous Prostration.

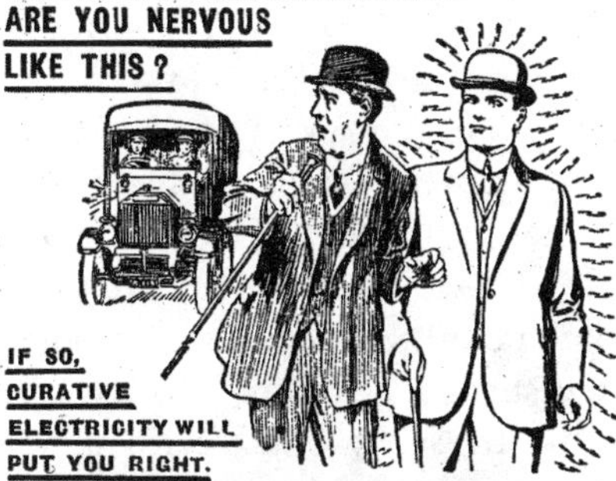

The Pulvermacher Appliances are the only inventions for the administration of curative electricity, endorsed by over fifty leading Doctors and by the official Academy of Medicine in Paris.

HAVE YOU ANY OF THESE SYMPTOMS?

Are you Nervous, Timid, or Indecisive?
Do you lack Self-Confidence?
Do you dread open or closed spaces?
Are you wanting in Will-Power?
Are you "fidgety," restless, or sleepless?
Do you blush or turn pale readily?
Do you shrink from strange company?
Are you subject to sudden impulses?
Do you crave for stimulants or drugs?

If so, you can safely assume that you are suffering from Neurasthenia. The neurasthenic also often suffer from **Indigestion, Liver Troubles, Constipation, Palpitation, Loss of Appetite, Excess of Appetite,** and a host of other disorders due to faulty functioning of various organs. Electricity is the only force that naturally supplies this deficiency of Nerve Force and restores tone to the whole nervous system. To-day you can be

CURED IN YOUR OWN HOME BY ELECTRICITY

by simply wearing the Pulvermacher Appliances, which are light, easy, and comfortable to wear. They give no shock, but all the time they are being worn they supply the nerve centres with a continuous flow of electricity, naturally stimulate the circulation of the blood, and increase nerve nutrition. This is the natural and physiological treatment of Neurasthenia, which drug treatments can never cure. The Pulvermacher Treatment has cured the most obstinate cases of Neurasthenia and Nervous Disorders when all other methods have failed. If you are suffering from any form of Nerve Trouble, or if you have any of the symptoms as described above, write to-day for a book that may well prove of incalculable health value to you; yet it costs you nothing. It is entitled "Guide to Health and Strength," and will be sent post free. **Those who can call personally are cordially invited to do so, when a consultation on their health trouble may be secured absolutely free of charge** and without obligation between the hours of 10 and 5.30 daily.

FREE COUPON

By posting this **FREE FORM TO-DAY** you will receive the "**Guide to Health and Strength.**" You place yourself under no obligation by applying for this book and particulars of the Pulvermacher Appliances.

Name...

Address...

....

☞ **Post to the Superintendent, Pulvermacher Electrological Institute, Ltd., 28, Vulcan House, 56, Ludgate Hill, London, E.C. 4.**

London Magazine. April, 1921.

6. Male curative electricity advertisement, 1921. *London Magazine*, vol. 46, no. 126 (April 1921): xiv. Copy in the author's collection.

a degenerate highbrow caste. It is narrated by a magazine editor named John Potifer, who awakes after a motor accident to find himself in a ruined, stinking place that he swiftly recognizes as the entrance hall of the British Museum. The rough-spoken inhabitants insist on calling him Alf, and they are greatly perturbed by his altered form of speech and sudden foolhardy determination to enter the forbidden library reading room. The disoriented Potifer/Alf feels as though his body is inhabited by "two separate beings," one "the real me" and the other "an illiterate savage."[32] In moments of peril, as when he is attacked by a pack of wild dogs, the savage takes over: "I ceased to think, drew my sabre and slashed out furiously at the leader of the pack, a big, yellow brute, who might have been a mastiff gone far astray."[33] Potifer's "local self" is also able to dodge spears before the "real me" even realizes there is danger. Discovering that only "Four Eyes" are permitted to enter the reading room, Potifer searches the overgrown vestiges of Bloomsbury for a ruined optician's shop, acquiring spectacles that invest him with a fearful power. He penetrates the reading room to discover a band of illiterates at an altar, burning their way through the collected works of civilization. Governed by a "Librarian" who pretends he can read, the Four Eyes demand food from those outside by threatening to prick them with an allegedly fatal syringe. Potifer kills the Librarian and evades another pack of wild dogs before regaining consciousness in a nursing home in his own time. The narrator of his tale dismisses as "pure bosh" the theory that Potifer's soul has traveled to another time period, preferring to believe that the motor accident has activated "some hitherto dormant brain cell" which "built up the airy fabric" of his yarn.[34] As with Mr. Fipkins three years later, it is safer to regard the experience as a dream, the product of a temporarily disturbed mind.

Again, the story's evolutionary frame recalls *The Time Machine*. Finding himself in a degenerate future, Potifer deduces that "the descendants of the more or less civilised folks I had known had completed the circle and returned to barbarism."[35] Yet his body is simultaneously inhabited by two men from different points on the circle: a savage named Alf (implying that caricatures of the East End may encapsulate London's future) and a magazine editor named John Potifer. These two identities combine to spawn a third, a misunderstood hero wearing spectacles who strikes fear into the heart of fellow savages and overthrows the tyrannical rule of pseudo-intellectuals. The blending of distinct evolutionary epochs gives this British Museum adventure a disruptive Einsteinian flavor, unsettling the Wellsian story arc and raising questions about what kind of man Potifer really is. But the narrator's verdict safely stows these multiple identities from different evolutionary moments in their rightful place, buried in the unconscious. Dream conclusions carry a suggestion that theories of time and space may conspire with psychoanalysis to render modern life excessively complicated, presenting individuals with more concurrent modes of existence than they can reasonably accommodate.

Female characters are scarce in these temporal adventures, but the *Red Magazine* did feature one woman with a taste for peeking into the future. Leslie Beresford, another pioneer of science fiction for the British pulp market, contributed "The Stranger from Somewhere" in August 1923. The setting is an isolated village close to the rugged Yorkshire coastline, where Hector Garden is grappling with the papers of his recently deceased uncle, a famous physicist. Amos, whose Semitic name casts him as a modern prophet, believed that "time does not exist at all" and claimed that the "light which travels through the ether illuminates at once the Crucifixion and the future of the Earth in, say, two thousand years of our time."[36] Hector finds this "highly intense matter" tough going and is relieved when his neighbor, a man named Rutherford, interrupts with an invitation to dinner. Ernest Rutherford's success in atomic disintegration had been reported in the *Daily Mail* during the week of Einstein's 1921 visit, and the name would have prepared readers for a story about dangerous powers.[37] Rutherford's sister Molly constitutes the story's love interest, and she, rather than her brother, is the main stimulus for scientific discussion. She is also the first of the three characters to be confronted with the story's strange visitor and to associate his appearance with Uncle Amos's Einsteinian theorizing.

The effort of preparing his uncle's manuscript, "The Future of Mathematical Exploration," for publication has led Hector to neglect his courtship of Molly, and he tries to make amends by explaining what he has gleaned from it so far.[38] Amos had been pursuing Professor Einstein's suggestion that "if anybody could be projected from the Earth into Space at the same speed as light travels and projected back again, they would not have aged a single second. Whereas, the Earth would have aged perhaps a thousand years or more."[39] Unlike male characters who tend to complain of headaches and confusion at the slightest hint of extra dimensions and light-speed travel, Molly is entranced. "How startling to come back to Earth, and find it all different," she muses, wishing that "one peep at the future" were possible.[40] But Hector is more interested in "a future not quite so distant," and he takes the opportunity to ask whether she will become his wife. Lovemaking is interrupted by the arrival of a meteor and the intrusion of a very thin, very tall figure clothed in "some kind of white flexible metal," with "blue eye-shades" and the ability to burn through metal shutters.[41] Described as a "man's girl, daring in sport, full of pluck," Molly is not given to swooning—there is clearly no call for a dose of curative electricity here.[42] But a lone confrontation with the white figure reduces this modern woman to a more old-fashioned feminine display of screaming and fainting. The fascination of light-speed travel has been dispelled in a chilling encounter with a future being that "wasn't . . . a bit human."[43] Confronted by the stranger in their turn, the two men find their minds taken over by incomprehensible sounds as they experience a vivid nightmare vision of the local area "blotted out by great far-stretching metal structures" and teeming with "queer people" like the white figure.[44] "Something happened to both their minds," the narrator explains, "which they afterwards could only compare with what you call 'jamming' in wireless operations."[45] Rutherford, who has the advantage of understanding Russian, is able to discern three words repeated over and over again: "Time," "Backwards" and "Light."[46]

Beresford's "China-white people," all wearing identical blue eyeshades, are much harder to comprehend and communicate with than savages in a crumbling British Museum. The Librarian's rule might have been based on deception and superstition, but Potifer could at least use reason to navigate this debased form of trusteeship. These uniform white beings are beyond the reasoning of Molly and the two men, populating the Yorkshire village's projected future space with nonhuman forms that look and sound as though they may be descended from Bolsheviks. Reluctant to follow Molly's belief in his uncle's theory, Hector concludes that it is better to forget the "outrageous piece of human mechanism." There is, after all, no evidence to prove "from where he had come, from this or any other world, from the past or the future." Whether time projection is possible or not, it is wiser to abstain from future visions of rural England—especially if they speak Russian.

A Vivid Mental Picture of the Queer Happening

Why was physical experience of the fourth dimension repeatedly confined to the supernatural realm in these stories? There were no dedicated science fiction magazines in Britain until well into the 1930s, but popular magazines accommodated plenty of science fiction stories during the initial period of Einstein sensation.[47] These stories were more common in the *Yellow*, which, as Mike Ashley notes, "syphoned off some of the more unusual stories, allowing the *Red* more space for romantic fiction."[48] Brisbane in particular was not shy of departing the bounds of terrestrial experience. Besides an array of more conventional romances, animal stories, sea adventures, and exotic encounters, his contributions to the *Red* and, especially, the *Yellow* during the first half of the 1920s included an attack from the moon, atomic disintegration, a trip to Mars, alien societies on Saturn and Mercury, a shrinking elixir, a cyclone machine, buoyant hydrogen clothing, and the radiological origins of the short skirt.[49] And yet it was in the *Red* that relativity themes found their earliest treatment in British magazine fiction. Beresford's "Stranger from Somewhere," with its light-speed traveler and glimpse of a futuristic cityscape, comes closest to pursuing the possibilities of Einstein-themed science fiction. But its ending uses a conventional device from tales of the uncanny, concluding that it is safer not to probe too deeply into these matters. Designation of the new time and space as a theme for supernatural treatment was partly a response to the persistent lack of an accessible exposition: by declining to explore relativity through science fiction, these authors were registering the impossibility of amateur involvement with Einstein's universe.

The closest Brisbane came to articulating popular experience of relativity through the conventions of science fiction was in "An Elementary Affair," published in November 1923—a month after "Mr. Fipkins." This story describes the invention of Albert Uddington, a maverick professor of physics who works on "light, and so forth."[50] His name suggests a blend of Albert Einstein with his leading British exponent Arthur Eddington (though Uddington is also the name of a Lanarkshire town). The professor has discovered a ray that turns organic matter into stone, and it is only a matter of time before this modern Medusa beam is directed at cats, goats, and a prying servant, all in the interests of scientific advancement. He wears, of course, a "disreputable topper," and his house is laced with dust and cobwebs.[51] Inevitably, the professor gets caught in the beam of his own ray. He finds that his legs have turned to crumbling stone, a condition that rapidly spreads through his entire body. Entrusted with crucial notebooks when the professor realizes that he cannot survive, the narrator burns them to prevent the invention being used for military purposes. Brisbane's story playfully translates the paralyzing difficulty and dry exposition associated with modern physical science into a literal effect: not unlike Einstein's account of his theory in the *Times*, Uddington's invention turns anybody who comes near it into stone, and they are soon reduced to crumbling dust. But the ray, that conventional device of science fiction, is indiscriminate in its effects. The social nuances of the Einstein sensation required the conventions of supernatural storytelling, in which peculiar effects are more dependent on individual circumstance and capacity than on a gadget or potion.

The moral in Einstein-themed adventures is less clear-cut than in straightforward science fiction from the same magazines. Stories featuring villainous inventions were commonly used to treat the theme of atomic power, which had already been widely popularized.[52] Brisbane's "The Almighty Atom," published in the *Red* for March 1922, loosely connotes Einstein's musical interests and his famous equation for converting mass into energy.[53] Toward the end of 1922, the novel *Atoms,* cowritten by novelist Trevor Wignall and science author Gordon Knox, was serialized in the *Yellow Magazine.*[54] A romance in which the plain man's common sense is seen to triumph over the machinations of a power-mad Greek and his ruthless German henchmen, it was republished in book form by Mills and Boon in 1923 and won praise from Ronald Ross for its combination of "the wholesome powder of physics" with "the jam of a stirring tale."[55] But while the notorious difficulty of relativity theory may have lent it villainous qualities, the new cosmology itself was hardly going to facilitate world domination or mass destruction. What iniquitous power could an understanding of Einstein actually confer on those capable of mastering it, aside from boring their friends to death over dinner? And yet there was definitely something pernicious about relativity. It caused robust men to disappear and made sporting young ladies scream and faint. A character at the mercy of disjointed time might fall through a crack and end up outside time altogether, more radically lost than any jungle explorer or seafarer; might find his body inhabited by a savage self from a different stage on the wheel of evolution; or might be confronted by hostile, inhuman forms with whom no communication was possible.

The most telling example of relativistic distress is found in what may be the first piece of British fiction to depict the Einstein sensation: "The Q-Ray," by A. E. Ashford. This story was published in July 1920 in the same issue of the *Red* as Brisbane's story about book-burning future savages. Although Ashford mentions neither relativity theory nor Einstein by name, the affinity with recent press coverage of this topic is clear from the start of his tale. The Q-ray is a discovery of the late Professor Robert Gregsby, who achieved fame for his "erudite" work not because his expensive book sold well, but because of "the controversy it aroused in scientific circles," which, "chancing to be taken up by the popular Press at a time when news was scarce, naturally intrigued the man in the street."[56] Two years before Percy Harris despaired of popular opinion about Einstein, Ashford's narrator anticipates his verdict almost word for word: "Undeterred by the consideration that they could not have read two consecutive pages of the book without a headache, people fell out with one another as to whether Professor Gregsby had made the most momentous discovery of the age, was mad, or merely an impudent charlatan." But where Harris felt obliged to put frustrated readers in their place, Ashford sets incomprehension of abstruse theory to work in the service of a conventional uncanny plot device: the opening declaration of skepticism. Readers are thereby primed for the narrator to uncover the truth.

Gregsby's controversial ray is said to have precisely the property that disappointed readers of *Conquest* came to demand in exchange for their shilling a month: it renders the time dimension as tangible and accessible as space. But Gregsby's claims are regarded as dubious, for he had been unable to "direct his ray upon the future"—and discern, for example, the winner of next year's Derby (the standard test of any time-travel device).[57] The ray had then been accidentally lost, so that nobody else could share in the professor's supposed glimpses of the past. The events in the story are set some years later, when the narrator encounters an old friend named Brampton who is living in Gregsby's old house. Some of the scientist's abandoned equipment is still lying around, but this on its own is not enough to reactivate the time dimension. What opens up the ray's power is the narrator's willingness to believe, his intuition, and his strong visual imagination—in other words, the ideal characteristics of a supernatural story reader. To these Ashford simply adds an isolated rural setting and several bottles of home-brewed ale. Under these conditions it would be somewhat uncanny if the fourth dimension did not manifest itself.

Brampton, who, like the narrator, is an artist, has to have the Gregsby story dragged out of him. He keeps apologizing for such a far-fetched, "tommy-rotten yarn," which involves the former occupant's claim that a rug thrown over an unsightly relic of the professor's equipment had dematerialized before his eyes.[58] Quaffing the excellent ale, Ashford's narrator finds himself experiencing "a vivid mental picture of the queer happening," and he insists on investigating the greenish, glowing metal dais in Gregsby's former laboratory. As they search for a means to activate its alleged power, the two men are appalled to see "a faint nebulous figure" appearing on the platform, a man "densifying" as he gains flesh and blood before them.[59] Brampton is rooted to the spot with horror, but the narrator experiences "an isolated streak of mentality in a fog of confusion."[60] A part of his mind works "vividly, preternaturally" to project images: he pictures the original story being related to Brampton, and the rug etherealizing.[61] In a "queer, inconsequent, altogether detached way,"

7. "The Q-Ray," by A. E. Ashford. *Red Magazine*, vol. 44, no. 263 (July 9, 1920): 201. The Bodleian Libraries, University of Oxford, Per.2561.e.5907.

he then witnesses the terrified man materializing, hands "thrust out" in a "gesture of appeal." Suddenly the narrator comprehends that Gregsby must have sent a man into the future, with the intention of bringing him back again to report on the experience. But something went wrong, and now the professor is dead. The realization comes too late, and the traveler dematerializes again before the two men can pull him off the dais.

Ashford's story is not concerned to engage with relativity on scientific terms: the narrator conceives of the lost traveler arriving "at that point where the two dimensions of Time and Space ever intersect, that pinpoint of Time which is ever-shifting forward, second by second, the immediate Present." But its movement from headaches caused by reading about Gregsby's work in the newspaper to the vivid sensory experience of contact with the ray's victim shows acute sensitivity to popular disaffection regarding access to Einstein's universe. The narrator's imaginative capacity gives him an instinctive understanding of the effects of Gregsby's ray. While his more skeptical companion is paralyzed with shock, his own "preternatural" vision projects past events (the rug disappearing; Brampton being told about it) and the present situation onto a kind of cinema screen in his mind. This is comparable to the way in which Leslie Beresford's male protagonists feel as though their minds have been "jammed" with wireless signals in a strange language. The authors of both stories recognize that readers of fiction and newspapers also go to picture palaces and tinker with radio sets, and that as consumers of these technologies, they have intuitive knowledge of disrupted time and space.[62] Features of Einstein's universe are understood in the context of modern media experience, including the "trick advertisement" suspected by Mr. Fipkins and the body belt advertisement in "The Time Adjuster."

The preternatural projections experienced by Ashford's narrator differ from the "cinemativity" playfully proposed by the *Daily Express* correspondent who cajoled Einstein into agreeing that the cinematograph might render "time and space as puppets of the screen," and from Professor Low's illustration of curved light rays by means of a "film of which the man in the sun was the producer."[63] The two artists in "The Q-Ray" are presented with something disturbing and distressing rather than educational or entertaining. Gregsby's invention also differs in its psychological impact from Uddington's ray and from the secrets sought by villains in atomic narratives. Contact with the Q-ray enhances the narrator's intuitive understanding of events while leaving him powerless to save the flickering, terrified figure of a lost man. The artist's imagination brings him closer to the scientist's accomplishment, only to yield a sharper sense of damage done. The Q-ray stands less for the pursuit of destructive powers than for paranormal insight compounded with a paralyzing inability to act on the knowledge obtained.

By incorporating relativity themes within uncanny narratives, pulp authors were able to accommodate a contradiction that had arisen through newspaper coverage of relativity: ordinary readers might know nothing about Einstein's theory, but they already knew plenty about disrupted time and space. Such knowledge came from multiple, intersecting sources, particularly experiences with new media and the recent war.[64] Flickering images of lost men had become a part of everyday life through wartime cinema newsreels, and later, wireless technology developed for military use entered the home.[65] The slaughter in Europe led to loss of faith in army authorities when millions of soldiers were sent to their deaths for questionable gain, and a backlash against propaganda revealed the extent to which those at home had been lied to through the mass media.[66] Even without these wartime connotations, the cinema and, later on, wireless, were the subject of diverging views about whose interests were being served. The issue of *Conquest* featuring John Ambrose Fleming's article "Some Difficulties in the Theory of Relativity" included an appeal from the editor for a dedicated "Popular Science Picture Theatre."[67] Harris regretted that "the general intellectual level" of moving pictures "is still very low, due to the 'showman' influence which still dominates the industry," and called for readers to demand the screening of neglected films on such topics as the Shackleton expedition, the cuckoo, the manufacture of steel, and the Derby finish.[68] The first BBC broadcast in November 1922 marked the arrival of a new force in the ether, one whose authorities were similarly determined to elevate listeners to a higher level of culture.[69] New media technologies were invading more minds, more intimately and spectacularly than ever before, but would the content be determined by mass audience tastes or by cultural elites?

In marking Einstein themes as supernatural, pulp authors treated the news sensation around relativity in a different spirit from that seen in the popular science magazines. Instead of seeking to untangle science from sensational headlines, pulp authors sought to demonstrate an intuitive grasp of disrupted space and time on the part of mathematically innocent consumers, identifying relativity as a symptom of modern mass media. The Einstein sensation was reenvisioned as mental invasion by a broadcast of cosmic proportions, one that perpetuated the loss and disorientation associated with war and offered limited scope for control on the part of viewers and listeners.[70] Einstein was not responsible for this vision, but the themes associated with his theory lent themselves exceptionally well to exploring and expressing it. Significantly, the class conflict over relativity exposition became muted in this process. Stories were finely balanced, championing the interests and abilities of readers whose access to elite institutions was minimal without going so far as to incite any real challenge to cultural authority. The figure of Potifer in "Thus Said Pel!" encapsulates this balancing act, locating its hero between sham highbrowism and savage anarchy. Reading these stories in their political context, we may detect the enrollment of Einstein's universe in the defusing of revolutionary sentiment, with four-dimensional space-time being depicted as a dangerous place that readers are equipped to access but would do well to avoid. The radical potential of these stories is at once enabled and stifled by the medium in which they appear: mass-market fiction knows what the mass of readers want, but must translate action into entertainment in order to stay in business. It is this fleeting paradox that distinguishes Einstein-themed stories in the early 1920s pulps from earlier and later tales of the occult in which perils of the fourth dimension are not so closely linked to representations of mass media.[71]

The Cold Chasms of Interstellar Space

In the mid-1920s the new space and time began to feature in romantic fiction, where genre conventions allowed the dangers of cosmic abstraction to be neutralized more decisively. Physics and astronomy were shown to be the opposite of successful intimacy and procreation: men who pursued astronomy, and women who fell in love with them, risked a lifetime of loneliness. A simple way to signal any character's isolation was to link them with the stars. The inhuman connotations of modern astronomy were nothing new—they can be seen, for example, in Walt Whitman's 1850s poem "When I Heard the Learn'd Astronomer." But the arrival of relativity provided a deeper form of abstraction through which to articulate the vulnerability of passion and intimacy in the modern world. In these stories, penned by men and women, it is the female characters whose commitment typically overcomes the pernicious effects of cosmology.

The *Novel Magazine*, with its emphasis on novelty and innovation in storytelling, was particularly alive to the ways in which science and technology could be used to give familiar romance themes a fresh twist.[72] The presence of radio in many homes, for example, suggested that affection and desire might behave like waves in space. Wireless communication played a different role here than in adventure stories, where it was generally used to establish contact with Mars or attain other enterprising goals.[73] In the romance genre, wireless tended to be allied with the triumph of domestic intimacy over obstacles threatening to disrupt family life. "A Wireless Uncle," in the *Novel* for May 1924, relates the difficulties of a young boy whose mother has a new husband.[74] It is only when the child identifies the stranger with a much-loved voice telling stories on the radio that a family relationship can be established. Wireless bonds were the opposite of astronomical coldness. The vignette for "Whispers on the Wind," contributed by H. W. Leggett to the *Novel* for August 1923, enticed readers with the promise of love conquering distance: "Between two people who love each other deeply there is sometimes a mysterious communion of the spirit, which rises above the actualities of time and space."[75] Joan's discovery that she is pregnant calls Michael home from abroad, where she has left him following difficulties in their new marriage: "Even across dividing seas this wireless of the heart reached out, and in the greatest crisis of their lives saved them from disaster." Attempting to make amends before his wife's departure, Michael had promised to get an astronomy book she wanted.[76] But stargazing fades from her mind at the prospect of raising a family together. "The Bell That Never Rang," contributed six months later by E. W. Morrison, describes "a sort of mental wireless" enabling certain individuals to "communicate with each other in moments of extremity."[77] John's desperate need for help as his wife struggles in childbirth conveys a mysterious signal to his old friend Margaret Adamson, a nurse who has retained an unrequited passion for him. Brushing aside the thought that she now has the power to remove her rival, Nurse Adamson quickly gets on with saving the child and its mother, realizing that John's love for them exceeds anything he ever felt for her. As she departs into the night, the narrator tells how "a few faint stars were reflected in the wet pavements. They gleamed like the eyes of a kind, blind man, giving light even as they were denied the gift of vision."[78] Caught between the stars and their reflected light, Margaret appears destined for a lonely existence, helping others to nurture what she herself must live without.

Astronomy's association with foreign travel and long nights dedicated to observation made it the antithesis of successful marital relations. In the same issue as "Whispers on the Wind," Lee Foster Hartman's story "The Altar of Destiny" depicts an astronomer observing a transit of Venus far from home. Faced with the savage sexuality of a native girl who seems to have uncanny knowledge of his family and fiancée, he is saved from an improper liaison by her tragic death.[79] "The Morning Star," by R. L. Dearden, in the *Novel* for January 1925, brought cosmic perils closer to home: "Professor Merriless knew all that there was to be known about astronomy; but there was one star which he had overlooked—the morning star of love."[80] Gilbert Merriless, a forty-year-old professor of astrophysics at the University of Warchester, has recently married Joy Loveday, "twenty-two, pretty, utterly and distractingly feminine." But their reading habits are woefully mismatched. He is "a profound thinker, a voracious reader of 'works,' and, in a sense, a philosopher," who is "more at home with a spectroscope than with a pretty young woman." Her "mental pabulum," by contrast, consists of "stories wherein grim, strong, silent heroes" may be relied on to "fling the fluffy, inconsequent, illogical little heroine across their stalwart knees" in order to administer the "sound spanking" that "she, womanlike, had been longing for since the first chapter." Engrossed in his work on "Relativity and the Nebular Hypothesis," Gilbert lets his appearance go to seed and addresses his wife using "a sort of modified baby-talk" which, the narrator observes, "would have been an insult to the intelligence of a child of twelve, let alone a desirable young woman."[81] Joy begs him to teach her astronomy so that she can help in his research, but he tells her to stick to the love stories, believing that the gulf of knowledge is "too wide and too profound to be bridged." What, he asked himself, "could this charming but abysmally ignorant child ever know of cosmogony or 'four dimensional space'?" And so she withdraws into a world of "silent he-men and self-willed *ingénues* who came to heel," her bookshelves lined with brightly colored jackets sporting titles like "A Cave-Man's Love Story," "Lochinvar in Plus-Fours," and "The Taming of Tiger-Cat Carstairs."[82] Enter Minerva Alston, "a talented lady of uncertain age who had graduated with high honours at the University of Cambridge," a "wrangler" whose regular teatime visits cause the professor and his wife to drift further and further apart.[83] The situation is saved at midnight on Christmas Eve when the sound of children singing diverts Gilbert Merriless from his manuscript on relativity. The simple carol is like a "wandering flock of angels" who have "bridged the cold chasms of interstellar space in order to pour for a moment their glad message of divine love and divine compassion into the ears of tired mortals."[84] Captivated by this angelic song, the professor becomes once more "a child in heart," approaching nearer "to Him whose power stretches beyond a million light-years, and a million suns." He rushes upstairs to find Joy mumbling about "cosmogony," "condensation," and "aggregation" in her sleep.[85] The garish book cover in her hand depicts "a girl in the arms of her cowboy lover," but underneath it lies a copy of "Merriless on Rational Cosmogony." Awakening, she confesses her Christmas surprise: "Oh, Gilbert, if it should be a son—I—I wanted him not to be ashamed of his ignorant little mother." Gazing up at Phosphorus, the morning star, they are reconciled.

Dearden gives readers the conventional ending they expect while teasing their tastes as consumers of romantic fiction. The clichéd book titles on Joy's bookshelves, with their bright covers and stock characters, are matched by an equally caricatured vision of science: the choice of names for the professor and his university and his falling out of touch with the simplicity of human contact. Yet the story also features authentic details such as wranglers, spectroscopes, and the nebular hypothesis. The only element of fantasy here is that Minerva Alston is ahead of her time in becoming a wrangler (a high-ranking candidate in the Cambridge Mathematical Tripos, an essential first step for a career in British astronomy). Women were allowed to sit Cambridge University examinations from the 1870s onward but were not awarded degrees until the 1940s.[86] Spectroscopy, in which light is passed through a prism and the spectrum analyzed, was a vital technique for unlocking the interiors of stars, enabling the new science of astrophysics to prove its value during the early 1900s.[87] The nebular hypothesis, first suggested at the end of the eighteenth century, proposed that the solar system was formed from a swirling ball of gas that condensed to form the sun and planets.[88] While relativity theory was not directly relevant to this idea, cosmologists continued to refine the nebular hypothesis during the twentieth century in relation to the latest models of the universe. These professional touches are drawn into the service of the romance plot, giving the professor's work a weight of seriousness in contrast to romantic fiction. The image of Joy Loveday, mumbling restlessly in her sleep with "Merriless on Rational Cosmogony" tucked inside "The Taming of Tiger Cat Carstairs," suggests that books on different topics may nestle inside one another more easily than individual readers can overcome the gulfs represented by their choice of "mental pabulum." Only when these characters set aside their books to concentrate on spiritual values and the nurturing of new life can the isolation caused by divergent reading tastes be resolved. Beneath the story's saccharine moral of Christian love lies a more subtle verdict about the undermining of social life by consumption of increasingly specialized subgenres of print.

A broadly similar romance plot had appeared the previous year in one of the more upmarket fiction magazines. The *Novel* was a bargain at only 9d. a month during the mid-1920s, and Mike Ashley describes it as "cheap enough to be within the grasp of domestic and manual workers, but with sufficient prestige, through its better known authors, to have a veneer of quality."[89] The American monthly *Harper's Magazine*, which reached a substantial European readership through a London edition, cost a shilling a month and was printed on glossy paper. It was expensive and cosmopolitan enough to give readers the satisfaction of a quality magazine without appearing too "highbrow" or ignoring popular taste. Susan Ertz was at the beginning of her writing career when she contributed "Relativity and Major Rooke" to *Harper's* in April 1924.[90] Her story foregrounded the magazine's transatlantic reading experience by creating a protagonist named Mrs. Harper, a widowed American magazine writer who meets a retired British major while researching articles in London. A tentative courtship develops, only to face disaster when Mrs. Harper invites Major Rooke to attend a lecture on space and time. After two hours of hearing about the age of the earth and the speed of light, Major Rooke is devastated by the revelation of his own insignificance in relation to the vast scale of the cosmos. He feels "completely crushed by Professor Brightman's universe" and firmly believes his worth in relation to Mrs. Harper to be like that of the earth in relation to the rest of space.[91] He returns to his rooms "more confirmed in his bachelorhood than ever, thanks to that vile and interesting lecture which had pulled all the new supports from under his self-esteem."

The crippling effect of cosmology is reversed when Mrs. Harper persuades the major to hear another lecture, titled "Present Day Psychology." Praising the "vast complexities" of the human mind, Mr. Reeves Smedley declares that humanity is about to experience "the dawn of new faculties, hitherto regarded as supernormal."[92] He too has found Professor Brightman's cosmic scale daunting, but he reassures the audience that its relevance is limited and offers them a quotation from one of the founders of American psychology as solace. In *The Varieties of Religious Experience* (1902), William James argued that "so long as we deal with the cosmic and the general, we deal only with the symbols of reality, but *as soon as we deal with private and personal phenomena as such, we deal with realities in the completest sense of the term.*"[93] Rescued from cosmic insignificance, Major Rooke catches the hand of his companion: "Their fingers locked, and as that private and personal phenomenon took place their faces turned slowly toward each other, and in the semidarkness their eyes met, and Major Rooke dealt with realities in the completest sense of the term."[94] Diverging from the uncanny adventure stories, in which disruptive cosmic imaginings are safely stowed in the unconscious, Ertz's story offers modern psychology—with its promise of supernormal new faculties—as a route to rediscovering personal significance. This message, and the story's conclusion, are not so very different from the "mysterious communion of the spirit, which rises above the actualities of time and space" found in pulp narratives of wireless affection.[95] But the quote from James would have given *Harper's* readers the sense that they were consuming something a touch more erudite. Readers who felt overwhelmed by the bewildering array of knowledge on offer would have enjoyed the indulgence of raw sentiment without feeling that they had descended to the level of purely sensational writing. Where the conventional ending of supernatural tales allowed authors to neutralize relativity's revolutionary associations, the required happy ending of romantic fiction simply erased questions of social difference in the upholding of family values.

If Our Minds Could Only Grasp It

By the late 1920s the class politics of relativity had shifted, and the conflict over mass access to elite knowledge had yielded to bourgeois ownership of the new time and space. The theme of bewilderment among the masses was replaced by caricatured images of the upper classes, who were

8. *Punch,* December 11, 1929. Reproduced with permission of Punch Ltd., www.punch.co.uk.

expected to remain as blissfully ignorant of the latest discoveries about the universe as they were about political developments threatening to undermine their status. A cartoon in *Punch* for December 1929 captured this vein of humor by depicting the inevitable Einstein conversation at an exclusive dinner party. A balding, bespectacled professor turns to his hostess during the soup course: "Have you any views on Relativity?" he inquires. "Well—ah—I daresay there is something to be said for it," comes the lady's reply, "but they'll nevah dare to enforce it in this country."[96] Six months later another cartoon depicted two svelte young ladies reclining on an enormous couch, one of them waving a cigarette holder. "One never hears anything of Einstein's theory of relativity nowadays," one of them remarks, "I suppose he's a bit *démodé.*"[97] Her companion's response rolls together ignorance of Darwin and Einstein: "Well, I never was very much interested in my ancestors."

These cartoons mark a fresh wave of interest in relativity themes occasioned by the arrival of best-selling expositions by Arthur Eddington and James Jeans. The publishing phenomenon represented by these books helped to change the class dynamics associated with conversation about Einstein. Two stories in *Strand Magazine* for 1934 showed household names in fiction playing out what had by now become a distinctly bourgeois experience of reading popular physics. The *Strand* had carried the first Sherlock Holmes short story in 1891, and the magazine's initial 60-year run to 1950 included authors well known today, such as Rudyard Kipling, Agatha Christie, G. K. Chesterton, and Graham Greene.[98] Vivid drawings and elegant title fonts, high-quality paper and the inclusion of engaging nonfiction articles increased the sense of affordable quality. P. G. Wodehouse had first contributed in 1905 when he was twenty-four years old, and each year from 1910 to 1940 the magazine included several of his much-loved stories about the idle rich. "The Amazing Hat Mystery," featuring Percy Wimbolt and Nelson Cork of the Drones Club, combines the fourth dimension's uncanny qualities with its detrimental effect on courtship.[99] "You know how things go," explains the bearer of this tale to a friend laid up in a nursing home with a broken leg.[100] "I mean to say, something rummy occurs and you consult some big-brained bird and he wags his head and says 'Ah! The Fourth Dimension!' " Wimbolt and Cork have been fitted out with new top hats from Bodmin of Vigo Street to aid them in winning over two young ladies. Both relationships turn frosty, however, when Elizabeth Bottsworth and Diana Punter accuse their respective companions of wearing hats that are ridiculously

"ONE NEVER HEARS ANYTHING OF EINSTEIN'S THEORY OF RELATIVITY NOWADAYS; I SUPPOSE HE'S A BIT *DÉMODÉ*."

"WELL, I WAS NEVER VERY MUCH INTERESTED IN MY ANCESTORS."

9. *Punch*, June 11, 1930. Reproduced with permission of Punch Ltd., www.punch.co.uk

out of proportion to the size of their heads. Since a bespoke hat from Bodmin cannot possibly be the wrong size, the men conclude that there must be something seriously the matter with both females. Departing from the club, Percy and Nelson each encounter the former object of the other's affections and are much consoled by approving comments about their immaculately fitted toppers. The couplings are accordingly reversed, but the mystery remains: how could the hats have appeared so monstrously wrong in the eyes of each man's original sweetheart? When a nurse ventures that perhaps the two hats, ordered on the same day, might have been mixed up by the delivery boy (who had been glimpsed, along with an accomplice, larking about with them in the street), the raconteur finds her suggestion "ingenious" but "a little far-fetched."[101] "No," he concludes, "I prefer to think the whole thing, as I say, has something to do with the Fourth Dimension. I am convinced that that is the true explanation, if our minds could only grasp it."

Wodehouse's tales of upper-class misadventure were particularly appealing to the *Strand* readership because they allowed contemporary themes like the independence of women and the advancement of the working classes to impinge on the timeless upper-class world of the Drones without destroying it. Two years later the *Strand* carried "Archibald and the Masses," and both stories were included in the Wodehouse short story collection *Young Men in Spats* (1936).[102] "This here Socialism," remarks one young man at the story's beginning, "You see a lot of that about nowadays. Seems to be all the go."[103] Archibald Mulliner's foray from W1 to Bottleton East in search of closer contact with the "martyred proletariat" ends in a punched face, lost watch, and bruised ego. Stories about the Drones Club, like the more famous Jeeves and Wooster narratives from which they were a spin-off, offered light-hearted relief from the changing sex relations and class structures that middle-class readers had to navigate. By the mid-1930s socialism and relativity were well-established themes, and Wodehouse used their apparent novelty to members of the Drones as a way of signaling a sheltered existence. Far from being revolutionary, the new cosmology offered refuge from other more disturbing changes in society. It was easier for a young aristocrat to muse on the peculiar physical effects of the fourth dimension than to confront the vulnerability of a Bodmin topper to a delivery boy's mimicry or a modern girl's unorthodox opinions.

A contrasting depiction of aristocratic life for middle-class readers emerged in the Lord Peter Wimsey stories of Dorothy L. Sayers. In contrast to Bertie Wooster and his compatriots at the Drones Club, Wimsey is possessed of a sharp intelligence to rival that of Wodehouse's notoriously "big-brained bird," Wooster's manservant Jeeves. A few months before "The Amazing Hat Mystery," the *Strand* carried a Wimsey story that opens with the aristocratic detective quoting directly from a well-known exposition of relativity. The story, "Absolutely Elsewhere," is set in The Lilacs, the country home of a moneylender named William Grimbold, who has been brutally stabbed while eating his dinner. Chief Inspector Parker begins by explaining that "all the obvious suspects were elsewhere at the time."[104] This simple observation provides a cue for Wimsey to demonstrate the idiosyncratic reasoning process that distinguished his approach from the deductive reasoning of Sherlock Holmes:

"What do you mean by 'elsewhere'?" demanded Wimsey, peevishly. Parker had hauled him down to Wapley, on the Great North Road, without his breakfast, and his temper had suffered. "Do you mean that they couldn't have reached the scene of the murder without travelling at over 186,000 miles a second? Because, if you don't mean that, they weren't absolutely elsewhere. They were only relatively and apparently elsewhere."

Parker's retort to this inconsequential line of questioning immediately identifies the source of Wimsey's relativistic musing. "For heaven's sake, don't go all Eddington," he replies. "Humanly speaking, they were elsewhere, and if we're going to nail one of them we shall have to do it without going into their Fitzgerald contractions and coefficients of spherical curvature." The story ends with Wimsey having apparently traveled at 186,000 miles a second in imitation of the murderer, vindicating his opening assertion. Sayers's integration of Eddington and Einstein in her depiction of Wimsey and other bachelor characters is the subject of chapter 5. First, however, chapter 4 explores the source of those three technical terms, "absolutely elsewhere," "FitzGerald contraction," and "coefficients of spherical curvature," in Arthur Eddington's best-seller *The Nature of the Physical World,* first published in 1928. Eddington's fiction-friendly expository style helped to release Einstein's universe from supernatural inexplicability—though it didn't quite succeed in making relativity easy.[105] Sensitive to the nuances of language, Eddington picked up the theme of shadows from press coverage of relativity and translated it into the heart of his idealist interpretation of physics. But for some readers this only intensified the class conflict around the new cosmology.

FOUR

A Lady on Neptune: Arthur Eddington's Talkative Universe

In April 1938, at his home in a small village five miles outside Oxford, fourteen-year-old Tony Fleming receives a letter from his mother. She writes regularly, though their contact has been limited since he was entrusted to foster care at the age of three weeks. John Anthony was born in 1924 to a single woman working as a copywriter at Bensons advertising agency. Nobody but Tony and cousin Ivy, his foster mother, know the truth. His biological mother has since quit advertising thanks to a successful career in writing detective fiction, which she has been able to pursue in earnest since her marriage in 1926. Tony has assumed her husband's second name, for to outward appearances "cousin" Dorothy Sayers and Mac Fleming have adopted him. The now famous creator of Lord Peter Wimsey, ever mindful of her son's education and imagination, urges him to read the works of Eddington, a household name in popular science. Tony might obtain *The Nature of the Physical World* (1928), in the Everyman edition of 1935, for 2s. A more recent book, *The Expanding Universe* (1933), was more expensive at 3s. 6d., while Eddington's latest work, *New Pathways in Science* (1935), was priced considerably higher at 10s. 6d. Eddington could also be heard on the wireless, speaking on topics such as "Eclipses of the Sun" or "Other Worlds."[1]

Eddington's easy inhabitation of Einstein's universe was received with good humor among his colleagues. A columnist in the monthly astronomy journal *Observatory* remarked in December 1919 that the "vain attempts of the reporters to apprehend exactly in what the revolution consists have been amusing, and would have been more so but for our own similar difficulties."[2] On the next page a cutting from the *Evening Standard* for November 8 was reproduced. It described a reporter's visit to the Royal Society in quest of a "popular explanation" of the new theory. "Scientists Caught Out" was one subheading: the RS Secretary had "rubbed a hand over a dome-like brow, and frankly admitted he was beaten." The journalist persisted in his quest:

> **SORROWFUL CONFESSIONS**
>
> A distinguished scientist was next seen. "I don't understand it at all," he said, wearily. "Don't mention my name."
>
> Another equally distinguished scientist said:—"Einstein says, I think, that the qualities of space are relative to their circumstances. Is that what you want?"
>
> "No. Can you put the theory in terms of plums and apples?"
>
> "It can't be done. The theory is terribly involved and mathematical, and—don't mention my name—I don't understand it."

A final section, headed "One Man Who Knows," revealed that "Professor Eddington, of Cambridge, claims to understand the theory so, until he consents to put it in schoolroom prose—Gott strafe Einstein."

A minor mystery attends the reproduction of this cutting in *Observatory*, for the original article is not to be found in archived copies of the *Standard* for that week nor in any of the other London evening papers. It is probable that the light-hearted piece was dropped for the late editions, though it may conceivably be a work of fiction. The Royal Society has always had three secretaries at any one time: Physical, Biological, and Foreign. In 1919 the Physical Secretary was James Jeans, who was engaged in fierce debate with Eddington over the sources of stellar energy and unlikely to respond in the manner described above. The *Observatory* columnist stated that the clipping had been sent in by a reader, with the inquiry, "Did you write or see the *Evening Standard* column, 'Newton put in the Shade'? If you did neither, let me recommend it to you for further publication." It was not unusual for scientific institutions to provide copy directly to the newspapers, as Charles Davidson's article in the *Express* (discussed in chapter 1) demonstrates. Whether "Newton Put in the Shade" was written by a journalist, by an astronomer for the *Standard*, or by an *Observatory* wag as a pastiche for the amusement of colleagues, it testifies to a level of complicity between the press and the scientific community. The adaptation of a German army slogan, *Gott strafe England* (God punish England), is particularly redolent of close networking between astronomers and journalists, exuding a combination of schoolroom humor with the anti-German sentiment that professional scientific publications shared with the daily press at this time.

During the 1914–1918 conflict it had become quite normal for British learned journals to print negative statements about the Teutonic mind and the impossibility of valuable knowledge or culture emerging from Germany. Astronomy journals were no exception. *Observatory* for September 1916 featured a comparison between "German science" and "Latin science" by M. G. Boccardi, who described the formalism through which typically German science "oppresses, paralyses and kills."[3] In the "absence of a proper critical mind," Boccardi declared, German scientists were "transforming science into a mechanism and scientists into automatic machines." This "impersonal apparatus" was contrasted with the work of "Latin scientists," who were "proud to place on their work their own personal imprint." Earlier that year, a letter from Eddington had urged readers of *Observatory* not to think of Germans as Huns, pirates, or baby killers, asserting that "the lines of latitude and longitude pay no regard to national boundaries."[4] This plea was met with the accusation that he was "using his preconceptions formed before the war, and his own shrinking from horrors, to help him in ignoring actual facts," and the *Observatory* respondent reminded readers that "babies have been killed in ways almost inconceivably brutal, and not as a mere individual excess, but as a part of the deliberate and declared policy of the German army."[5]

Matthew Stanley, in his study of Eddington as a Quaker astronomer, has explored this exchange in detail because it illustrates a commitment to pacifism and internationalism in Eddington's professional life.[6] The astronomer manipulated existing publicity networks to promote the 1919 eclipse results as a triumph of internationalist science over wartime prejudice. He wanted Einstein to be proved right not just because the new theory represented an advance in knowledge of gravitation, but because the story of British astronomers testing a German theory could herald a new era of postwar collaboration.[7] This is a good example of how the "enrollment" described by Cooter and Pumfrey may unfold in practice.[8] During the late nineteenth century, astronomers had enrolled public audiences in a network of alliances that supported both astronomy and the British Empire by providing regular newspaper reports of British eclipse expeditions sent to far-flung corners of the world.[9] Readers were assured that money spent on telescopes and expeditions across colonial territory helped to ensure Britain's continued prestige as an imperial power. In the context of war in Europe, Eddington was able to exploit this network to a rather different end, enrolling audiences in a Quaker commitment to seeing Germans as fellow humans.

Eddington was skilled at adapting his language and tone to suit the needs of different audiences. The same stories are repeated throughout his expository writings, but they shift depending on the context. "The Domain of Physical Science," a chapter contributed by Eddington to *Science, Religion and Reality* (1925), rehearses key themes found later in *The Nature of the Physical World*, but the whole frame of the earlier chapter is quite different, making reference to "the man in the street," for instance, and using "we" for the expositor and "he" for the reader, instead of "I" and "you." This was not inexperience on the expositor's part: Eddington's rhetorical strategies for striking up a bond with mathematical innocents are in evidence in two articles he wrote for general intellectual periodicals in 1919 and 1920.[10] *The Nature of the Physical World* was based on a series of Gifford Lectures that Eddington had delivered at Edinburgh during 1927. The chief duty of a Gifford lecturer was to discuss the relationship between knowledge of the physical world and the existence of God, and this was an ideal opportunity for the Quaker astronomer to elaborate in a more popular form the special relationship between mind and matter that he had been formulating since the early years of his work on relativity. A closer inspection of Eddington's literary style and its development from his early scientific lectures onward helps to pinpoint what was "popular" about his writing and what it offered to middle-class readers between the wars. The analogies and stories that he developed for the Gifford audience may be contrasted with earlier versions from his Romanes Lecture, delivered at Oxford University in 1922 and published as *The Theory of Relativity and Its Influence on Scientific Thought*. By tracking the development of Eddington's expository arguments throughout the 1920s, we can see how spiritual commitments were incorporated into his account of relativity.[11] The affinity with popular fiction that emerges from this analysis highlights features that would have enabled readers to appreciate Eddington's writing for its relevance to their own experience, regardless of their religious beliefs.

Oiling the Reader's Mental Machinery

The text of Eddington's Romanes Lecture, delivered in Oxford's Sheldonian Theatre in May 1922, is the closest we can get to a "neutral" nonmathematical description of relativity for a British audience from somebody who was closely involved with Einstein's theory at an early stage.[12] Perhaps the most telling endorsement comes from Bertrand Russell's having adopted its key points as the basis for his *ABC of Relativity* (1925). Russell had developed his own philosophy in antagonism to idealist views, and he thought that religion was largely harmful superstition. So any hint of Eddington's later inclination to argue that the spiritual world and the world of physical law were alike based on "hard facts of experience" would have led Russell to fall out with the Quaker astronomer's approach much earlier than he did (the break came swiftly on the heels of *The Nature of the Physical World*).[13] In *The ABC of Relativity*, Russell was happy to adopt key arguments found in the Romanes Lecture, weaving them into anecdotes that accorded with his own, rather different, philosophical and political convictions.

The Romanes Lecture saw Eddington devoting all his skill as a storyteller and performer to dramatizing disputes between observers with different points of view. The first challenge he faced was to convince the audience that lengths contract along the direction of their movement. Contraction cannot be detected by those who travel with the thing being measured, and Eddington thrust his arm into the Oxford air to illustrate this conspiracy. Inviting his audience to imagine that the earth was moving at 8,000 times its usual speed around the sun in the direction of his arm, so that its length of one foot became just six inches, he challenged them to measure it. He then explained that they could trust neither the yard-measure nor the anatomy of their own eyes because every possible measuring device was also contracted by half along the direction of supposed motion.[14]

By 1928 Eddington had augmented this introductory feature, scripting a more elaborate and obstinate response on the part of his imagined reader or listener: "'Very well,' you reply, 'I will not get up. I will lie in bed and watch you go through your performance in an inclined mirror. Then my retina will be all right, but I know I shall still see no contraction.'"[15] The answer, of course, is that the mirror's own motion with the earth "introduces a distortion which just conceals the contraction of my arm." Eddington incorporates the usual objections from common physical experience into his script, and by giving a voice (that he can control) to these objections, makes that familiar tension between the expositor's special knowledge and the audience's common perceptions an essential part of his "performance." Gillian Beer has stressed the significance of this technique: "Instead of the privileged anonymity of the single scientific observer," she notes, Eddington "implicates himself and reader, foregrounding the I-you relation whose relations are never fixed. . . .

The reader is given a voice, interrupting, questioning, not always satisfied by Eddington's responses. This move is not persuasive only; it is an enactment of the theory he propounds."[16] Physical frames of reference are anthropomorphized, and habitual protests against the abstraction of physics are woven into narrative tensions that the storyteller must resolve.

In his Romanes Lecture Eddington enlisted two famous characters to help make the contraction hypothesis more convincing. Copernicus berates Ptolemy for not realizing that a circle drawn on earth is really an ellipse when viewed from his own vantage point on the sun. But the lecturer insists that "Ptolemy has a right to be heard," and he eventually concludes that both characters are correct.[17] This reconciliation between the ancient Greek philosopher and the Polish Renaissance astronomer cleverly dispels any anxiety that members of the audience may experience in trying to comprehend the FitzGerald contraction: tension between the characters is heightened, diverting energy from potential conflict between expositor and audience. By 1928 Ptolemy and Copernicus had been replaced by a team of physicists on the earth arguing with a colony of physicists working on a spiral nebula traveling at 1,000 miles a second relative to the earth: " 'A thousand miles a second!' exclaim the nebular physicists, 'How unfortunate for the poor physicists on the Earth! The FitzGerald contraction will be quite appreciable, and all their measures with scales will be seriously wrong. What a weird system of laws of Nature they will have deduced, if they have overlooked this correction!' "[18] As before, anxieties arising from the attempt to understand the disagreement are displaced by the injunction to be open-minded, to accept both views as correct. But Eddington had gained confidence in his handling of common-sense objections, and this episode is augmented by direct confrontation with an imagined listener. Eddington anticipates that some members of his audience might object to the introduction of "imaginary beings" and abruptly responds that they "must face the alternative of following the argument with mathematical symbols."[19]

In *The ABC of Relativity*, Russell was at pains to distinguish between measuring equipment and observers, noting that the term "observer" in discussions of relativity was apt to be misleading, making people think that human beings and their minds were involved when in fact the thing doing the observing was "just as likely to be a photographic plate or a clock."[20] Relativity was concerned, Russell stressed, with a "*physical* subjectivity, which would exist equally if there were no such things as minds or senses in the world."[21] Eddington wishes to convey the opposite message: that the physical world is dependent on activity of mind.

His nebular observer, he explains, is "subject to a common failing of human nature" in believing that the universe has been created with his own planet in mind.[22] The usual injunction to get beyond parochialism takes on a more confrontational tone at this point: "Hence he is (like my reader perhaps) disinclined to take seriously the views of location of those people who are so misguided as to move at 1,000 miles a second relatively to his parish pump." A profound move is quietly contained within those parentheses, placing the reader's anticipated rejection of imaginary beings on a par with the nebular physicists' parochialism. As Beer observers, Eddington's storytelling is "an enactment of the theory he propounds." The physical insight that all frames of reference have equal validity is tacitly extended to an epistemological assertion: mathematics and imagination are offered as equally valid routes to the same knowledge. This was precisely the view articulated through Einstein-themed uncanny adventure stories in the pulps, and Eddington also shares their verdict: that the new cosmology is all in the mind. But where mass-market fiction authors of the early 1920s used this verdict to warn readers away from close contact with Einstein's universe, Eddington uses it to establish greater intimacy.

Eddington was a keen reader of contemporary fiction and literary classics, and his journal records the consumption of detective fiction by Margery Allingham, G. K. Chesterton, Arthur Conan Doyle, Agatha Christie, and Dorothy Sayers alongside Jane Austen and Charles Dickens.[23] His relish for suspense and puzzle solving found an outlet in the snappy exchanges that he scripted around disputes between different frames of reference. The Romanes Lecture features a β particle, emitted from a radioactive substance, taking issue with a scientist who marvels at its astonishing speed of 100,000 miles a second: "'That,' says the electron, 'is a matter of opinion. So far as I am aware I am at rest, if the word "rest" has any meaning. In fact I was just contemplating with amazement *your* extraordinary speed of 100,000 miles a second with which you are shooting past me.'"[24] As usual, it turns out that neither one has a stronger claim to being at rest. These examples were devised to illustrate Einstein's special theory of relativity, which is restricted to frames of reference that do not speed up or slow down while arguing with one another. It was much harder to give readers a purchase on the general theory of relativity, where the same principle is applied to observers with changing velocities (including bodies subject to gravitation).

For the Romanes audience Eddington approached the new theory of gravitation by staging an argument between Isaac Newton and an apple dropped by a man who has tumbled out of an airplane. From the free-falling point of view, Newtonian gravitation has been evaded: the apple cannot fall any faster than the hand that released it, and it appears to hover in midair.[25] The impossibility of deciding whether the free-falling apple or the earthbound Newton is correct helps to convey the need for a revised picture of gravitation that can accommodate both perspectives. In developing this altercation for *The Nature of the Physical World*, Eddington endowed the apple with more personality. He explained that, thanks to the "usual egotism of an observer," it has "deemed itself to be at rest," and looking down, has seen "various terrestrial objects including Newton rushing upwards with accelerated velocity to meet it."[26] Dismissing the prospect that it might invent "a mysterious agency or tug to account for their conduct," he turns the legendary fruit into a persuasive physicist. The apple points out that "Newton is being hammered by the molecules of the ground beneath him," and it displays a stronger grasp of physics than the famous mathematician: "Newton had to postulate a mysterious invisible force pulling the apple down; the apple can point to an evident cause propelling Newton up." Conceding that the apple has an unfair advantage here, Eddington then moves Newton to the center of the earth, where there is no gravitation and no hammering. The dispute continues, and this time the apple "cannot attribute Newton's acceleration to any evident hammering. It also has to invent a mysterious tug acting on Newton."[27] The impossibility of deciding which of them is really being tugged leads to the abandonment of gravitation as a force. Instead, it is identified with the curved geometry of the four-dimensional world.

This geometry is the hardest Einsteinian concept to make real without mathematics, and Eddington knew that it was impossible for his readers to visualize it. One strategy, which he adopted in both expositions, was to remind his audience that people had once believed in a flat earth.[28] Trying to fit a curved world into a flat frame of space-time was like trying to represent the earth's surface on a flat map. Once curvature was accepted, the distortions and complications involved in the flat picture disappeared. But this argument on its own does not produce a satisfying reading experience, for readers cannot simply be reasoned out of familiar conceptions.[29] *The Nature of the Physical World* makes the argument about intellectual progress more compelling by associating it with a higher level of active consciousness, more dialogue, and a sharper moral sense around the theme of parochialism. The apple has acquired egotism and the ability to look down, point out causes, and invent forces. Similarly, the speeding β particle is shown in the later version "smugly thinking itself at rest," wondering aloud "what it feels like" for the scientist to be moving so quickly and concluding that "it is no business of mine."[30] As

Beer explains, Eddington's expository technique is "not a matter simply of cajoling the amateur audience with personifications and invocations that will draw the audience persuasively into the unfamiliar discourse of physics."[31] Eddington does not stop being a scientist in order to become a popularizer. *The Nature of the Physical World* represents "a purposive, hard-working attempt by a physicist to establish an epistemology that will not reinstate the absolute, the out-there, in its language even as it attempts to destabilize them in its argument." Storytelling and philosophy of science are knit together: Eddington's "theoretical stance demands that he foreground human activity of mind in his description."

It was this thoroughgoing relativization that suited the "melting pot" philosophy at *Armchair Science* so well, earning high praise from Moore-Brabazon in his article "Physics and the Ordinary Man" (August 1930). Moore-Brabazon praised Eddington's refusal "to take a material view of life," his "gift of making extremely complex physical thoughts understandable to the man in the street," and the "vein of humour which runs through his works."[32] Eddington's concern to limit the scope of physical law, making room for spiritual experience, fitted in very well with the theory of knowledge expounded by Moore-Brabazon and Low, in which the constant change and fluidity of scientific disciplines were taken as a sign that the man in the street was qualified to engage with the experts on equal terms. The reviewer for *Discovery*, Victor Pullin, was also enthusiastic, welcoming the book's harmony of science and religion and praising the expositor "whose humility expresses itself in sympathetic help and quick humour which will be found to be remarkably efficient in oiling the reader's mental machinery when the subject matter becomes difficult."[33]

A rather different line was taken by critics who worried about the effect of Eddington's exposition on readers who were not equipped to work out for themselves where scientific exposition ended and religious values began. Eddington's name was frequently yoked with that of fellow popularizer James Jeans in this regard.[34] The Liberal politician Herbert Samuel regretted the "vague notion, emanating from these discussions and spreading among the public at large, that science does not know where she stands," adding that "the perverted conclusion is sometimes drawn that it has become legitimate to believe or not to believe anything that you please; any ancient superstition or the most irrational myths of a discarded theology may now be re-established in credit with the acquiescence of science."[35] Lizzie Susan Stebbing, a philosophy professor at the University of London, was also dismayed by the popularity of books by Eddington and Jeans. Her book *Philosophy and the Physicists* (1937) was reissued in the Pelican series in 1944, where it could be found alongside Jeans's *The Mysterious Universe*. Jeans's book was first published in 1930 by Cambridge University Press, and Michael Whitworth has shown that it was marketed on the back of Eddington's success with *The Nature of the Physical World*. This meant that criticism of Jeans also affected the reception of Eddington's book: "As Jeans's reputation sank, it dragged down that of Eddington."[36] Stebbing's book, she explained, was aimed at philosophers and "that section of the reading public who buy in large quantities and, no doubt, devour with great earnestness the popular books written by scientists for their enlightenment."[37] She found both authors guilty of "an amount of personification and metaphor that reduces them to the level of revivalist preachers," and she sought to expose the "emotional fog" with which they surrounded their subject matter.[38] These two "Alarming Astronomers" were found guilty of misleading the "common reader," persuading him "that there is a God who has created the world, who has designed man as the crown of this creation, and who will thus not leave him uncomforted; that Reality is spiritual; finally, that human beings can determine their own destiny."[39] Where Moore-Brabazon celebrated the empowerment of readers in Eddington's mysticism, Stebbing identified the professor as a charlatan overstepping his authority as a scientist. But Eddington's readers were concerned with other matters besides the relations between science and religion. The urgency of that topic can obscure other levels on which Eddington's expository devices would have resonated with the texture of everyday life between the wars.

Absolutely Elsewhere and the FitzGerald Contraction

Eddington began his discussion of time in *The Nature of the Physical World* by encouraging readers to "protest in the name of commonsense against a mixing of time and space."[40] He then invited them to accompany him on a journey "into the virginal four-dimensional world," using the first person plural to promise that "we will carve it anew on a plan which keeps them entirely distinct."[41] It is on this journey that Lord Peter Wimsey's "absolute elsewhere" crops up, in a series of diagrams representing successive stages toward the new plan for separating time and space. Describing himself as "a kind of four-dimensional worm," the professor begins by drawing a vertical line to represent his own endurance from the past (at the bottom of the page) to the future (at the top).[42] A horizontal line, crossing his worm-self at the moment Now, represents all the

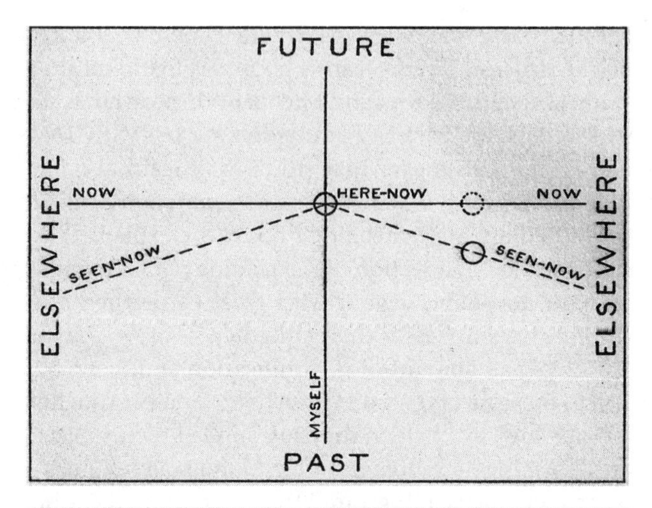

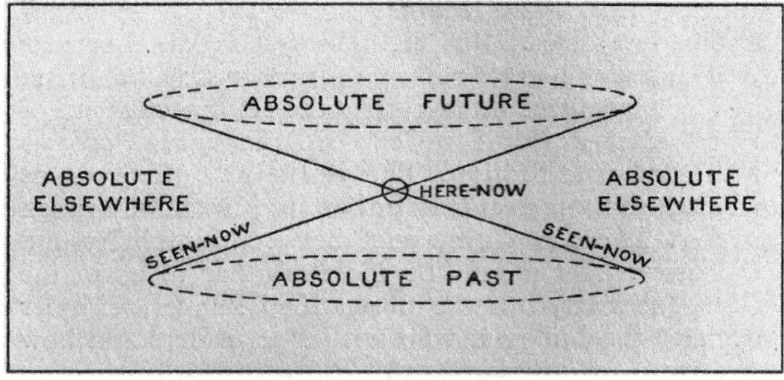

10. Arthur Eddington, *The Nature of the Physical World,* chap. 3, "Time." Reproduced by kind permission of The Master and Fellows of Trinity College Cambridge.

events happening Now in other places. The intersection is designated "Here-Now." Two dotted lines, sloping downward to the left and right, signify events that have already happened elsewhere but are "Seen-Now" because of the time taken for light to reach the observer at Here-Now (see fig. 10, *top*). The purpose of these diagrams and the accompanying narrative is to establish that two observers, temporarily occupying the same place at the same time but moving at different velocities, will disagree about what is happening Now. There is, however, one thing that all observers will agree on: what is Seen-Now.

In Eddington's final diagram the shared Seen-Now lines form two cones, one containing events that are absolutely in the future for all observers from Here-Now, the other containing absolutely past events (see fig. 10, *bottom*).[43] Any events requiring a faster-than-light journey from (or to) Here-Now are said to be Absolutely Elsewhere and are consigned to the area lying between these two cones, a "wedge-shaped neutral zone which (absolutely) is neither past nor future." Here is the true separation of time and space that the professor had promised to deliver: "Events can stand to us in a temporal relation (absolutely past or future) or a spatial relation (absolutely elsewhere), but not in both."[44] Appreciating that by this point his readers would have had enough abstraction, Eddington reintroduces human communication as a way of bringing this new picture to life. "Suppose," he suggests, "that you are in love with a lady on Neptune and that she returns the sentiment."[45] A familiar strategy of separated lovers is used to demonstrate the vulnerability of romance in four dimensions: "It will be some consolation for the melancholy separation if you can say to yourself at some—possibly prearranged—moment, 'She is thinking of me now.' Unfortunately a difficulty has arisen because we have had to abolish Now. There is no absolute Now, but only the various relative Nows differing according to the reckoning of different observers and covering the whole neutral wedge which at the distance of Neptune is about eight hours thick. She will have to think of you continuously for eight hours on end in order to circumvent the ambiguity of 'Now.'" The lady on Neptune is Eddington's only female character, and it is she, like the women who overcome the deleterious effects of modern cosmology in romantic fiction, who makes the effort that sustains intimacy through a daily eight-hour vigil. During the 1920s this story of separation must have been all too close to the truth for the many readers who had lost a lover or close relative in the war. Eddington's conclusion acquires an unexpectedly somber tone: "From this point of view," he notes, "the 'nowness' of an event is like a shadow cast by it into space, and the longer the event the farther will the umbra of the shadow extend." As the optimism of the 1920s yielded to the prospect of another war, it would have been hard not to wonder how far into the absolute future the shadow of conflict in Europe must extend.

Wartime experience had made itself felt in less poetic terms right from the start of Eddington's involvement with publicity for relativity. In an address to the British Astronomical Association on November 27, 1918, he had used rifle bullets and sugar (rationed for three years, until November 1920) to give the audience of amateur astronomers a sense of the practical issues involved in testing Einstein's theory. The BAA had been founded by in 1890 Walter Maunder, who was inspired by his disaffection with the rigidity of astronomical work at Greenwich and the elitism of the Royal Astronomical Society.[46] Established to recognize and support the contributions of amateurs to British astronomy, the organization welcomed women as members from its inception and included them on eclipse expeditions.[47] The BAA's first postwar meeting, held seventeen days after the Armistice, proceeded much like any other. Maunder began by reading out a report by one H. M. Johnson on the eclipse of June 1918.[48] He was followed by Instructor Commander M. A. Ainslie, who showed slides of the 100-inch reflecting telescope being hauled up Mount Wilson in California. Its four-and-a-half-ton mirror had been manufactured in France and shipped to America just before the outbreak of war, escaping German destruction of the works at Saint-Gobain. The instrument's moving parts weighed over 100 tons, and their transportation up the mountain was no small feat. A worm wheel seventeen feet across had its 1,440 teeth ground in situ, forming part of an enormous clock mechanism that would drive the telescope and keep its focus steady while the stars made their nightly journey across the sky.[49] Members of the BAA did not have access to telescopes a tenth the size of the Mount Wilson monster, but they knew enough about driving clocks and silvered lenses to appreciate its magnificence. "The Hooker," named after the American hardware millionaire and amateur astronomer who funded its mirror, remained the largest telescope in the world until a 200-inch telescope was assembled at Palomar in 1948. Edwin Hubble used the Hooker during the early 1920s to help settle the question of island universes and later to find evidence for an expanding universe.[50] Back in 1918, as the age of modern cosmology was dawning, Commander Ainslie's lantern slides took the audience of British amateurs, gathered in the Gothic edifice of Sion College on London's Victoria Embankment, to the frontier of astronomical advances.

Professor Eddington followed with an account of the two expeditions being organized to observe the total solar eclipse of May 1919. Before describing the experiment to weigh light, he took a moment to remind his audience of the difference between weight and mass: "When we buy a pound of sugar," he observed, "it is not the weight but the mass that interests us; the same sugar weighs more at the pole than at the equator, but it will not go any further."[51] A body's mass, as the assembled astronomers well knew, "is measured by the force required to stop it when it is moving with known velocity." We know that light has mass because a "minute force" is required to stop a beam of light. But a "special experiment" was needed to settle the separate question of whether light has weight.[52] Turning from sugar to rifle bullets, Eddington then pointed out the limitations of Earth as a laboratory. We can see that a bullet has weight from its curved trajectory: it falls 16 feet from its original course in one second and 64 feet in two seconds.[53] But light moves at 186,000 miles a second, and a drop of 16 feet in 186,000 miles is too small to detect on the terrestrial scale. Besides, a ray of light passing close to the surface of the earth will be out of range in much less than a second. "There is only one body in the Solar System big enough to produce a measurable deflection of light by its attraction," Eddington declared, "and that is the Sun." A ray of starlight grazing the sun would, if light had weight, be bent, and the star would appear displaced as a result.

It so happened that the sun was due to appear amid bright stars—the constellation Hyades—during totality the following May, presenting an ideal opportunity for such an experiment. Eddington allowed three possible outcomes: no displacement at all (which would be "startling" indeed); a displacement of the amount to be expected according to Newtonian theory; and a displacement of double that amount, confirming Einstein's prediction. Without attempting a full explanation, the professor simply informed his audience that the new theory "assigns a lower speed of light in a gravitational field, so that the Sun's field will act like a converging lens." There was no talk of the fourth dimension here, and this description of the sun's gravitational field would have reassured any of the assembled astronomers who happened to catch a whiff of continental metaphysics emanating from the new theory. They knew where they were with lenses. To reassure his audience further, Eddington concluded with another homely illustration: "There is an appalling amount of light in an ounce," he observed, adding that "the cost of light supplied by gas and electric light companies works out at something like £10,000,000 an ounce." Here was the same analogy that had appeared nine months earlier in Eddington's Friday Evening Discourse at the Royal Institution and would reemerge a year later under Charles Davidson's name in the *Daily Express*. Like their professional counterparts, the BAA audience would have regarded abstract theorizing about space and time with suspicion. They were more at home with observation and empirical data, and the ounce of light was an amusing way for the professor to acknowledge that the ultimate goal of these expeditions was measurement. Davidson was also present at the meeting, and the junior astronomer followed Eddington's presentation with a brief account of travel options in Brazil (a choice between pack mules and a three-week steamboat journey) and the prospect of fine weather for the morning of the observation.[54]

The two teams of astronomers charged with weighing light departed by steamship from Liverpool on March 7, 1919. Eddington and his assistant Edwin Cottingham returned from Principe on July 14, while Crommelin and Davidson did not get back from Brazil until much later. The voyage to Principe was approximately 4,700 miles, via Lisbon, Madeira, and St. Vincent.[55] During the trip Eddington wrote nine letters to his mother and his sister Winifred, with whom he lived on the grounds of the Cambridge Observatory, and they also wrote to him with news from home. These letters, bridging the gap between life at home and life on the four-month expedition, bring a domestic angle to the grand experiment of weighing light. "It seems ages since I started off in a rush in the taxi from the Observatory," Eddington wrote to Winifred from Principe on May 5. "I do not know what has been happening to you for a long while," he continued, "indeed I do not know what has been happening in the world in general—whether peace has been signed or any important events have occurred."[56] At times the astronomers must have felt that they were living in a parallel universe untouched by the recent conflict. Matters of food, in particular, underline their distance from home. "I wonder if you are still rationed," Eddington wrote. "It seemed funny on the boat at starting to see full sugar-basins, unlimited butter, and to eat in a day about as much meat as would have been a week's ration." There were less welcome changes to bear: milk aboard the *Portugal* was not good, and Eddington had learned to take his tea black following "the usual Portuguese custom." He also had to adjust himself to foreign mealtimes: "coffee and biscuits in the cabin about 7 o'clock, déjeuner at 11, tea at 3.30, dinner at 6, and tea again at 9.30."[57] The déjeuner and dinner were "good meals to which I do justice, but the tea is very poor." Practically every letter mentions a variety of fruits, from bananas to pawpaws, and Eddington reported having consumed a dozen bananas a day during the stopover at Funchal.[58] But nothing can beat the English strawberries, and he repeatedly wonders whether he will be home in time to savor them.[59]

There are glimpses in the letters of the astronomer's international spirit at work, including his participation in a Good Friday egg and spoon race with Portuguese passengers bound for Principe, his reading of *The Vicar of Wakefield* in Portuguese to learn the language, and his games of tennis with the Curador and the Judge at Principe.[60] The energy exerted on these occasions, as on Eddington's long solitary walks, was typical of his active temperament but must also have helped to allay nervous anticipation of the momentous test. The last letter to his mother includes a nail-biting account of the eclipse itself, which was preceded by "a tremendous rain-storm." The astronomers caught glimpses of totality through cloud cover that "interfered very much with the star-images" and they acquired photographic plates in the grim hope that these would capture the necessary evidence (the status of these plates as "proof" of Einstein's theory has been a subject of intense debate by historians of science).[61]

Eddington A4/8

I wonder if you are still rationed. It seemed funny on the boat at starting to see full sugar-basins, unlimited butter, and to eat in a day about as much meat as would have been a week's ration. We have had no scarcity of anything since we started. I have, however, scarcely tasted ham or bacon (eggs have been plentiful) The milk was not good on the Portugal, and I have got into the habit of taking tea without milk, which is the usual Portuguese custom & is probably better in hot climates. I cannot get any swimming here, because of the sharks.

There are several dogs about here, one of them rather a nice terrier; but for the most part they are not up to much. At Madeira. [Nipper the dog at the hotel attached himself to me very much and followed me almost everywhere, although I did not encourage him at all, as he was neither beautiful nor free from fleas. He used to like to come and spend hours hunting lizards whilst we bathed.]

It gets dark here about 6 o'clock, and as one does not sit much inside the house, one does

11. Letter from A. S. Eddington to Winifred Eddington, May 5, 1919, A4/8. Reproduced by kind permission of The Master and Fellows of Trinity College Cambridge.

In this letter, penned on June 21 aboard the S.S. *Zaire* from Principe to Lisbon, Eddington acknowledged that he would very likely arrive home before the mail reached his mother and sister. The primary purpose of these letters, then, was not to maintain direct communication during his extended absence. His mother's letter (dated March 28) had arrived on May 11 and since then no more mail from Europe had reached him. In fact, as he explained at the start of his homecoming missive, "we expect to pass tomorrow (at S. Vincent) the outward boat which will be taking the next batch of letters to Principe."

The image of writing a letter to somebody whose most recent correspondence is steaming past as you write, on its way to the place you have just left, is an apt reminder of how slowly communication could unfold even in the modern age of telegraphs and trains. It is also a sign of how normal it was for people to engage in knowingly disrupted written exchanges, maintaining a connection even though evidence for this might not reach the other person until after they had been physically reunited. The war made commonplace the act of writing letters to soldiers at the front in the hope that they would be alive to read them, and the experience of receiving them at home after the writer had been killed. The telegraph and telephone enabled people to miss each other more frequently and to connect through minimized language and partial presence. "The voice that comes to us over the telephone wire is not the whole of what is at the end of the wire," Eddington noted, finding this a useful metaphor for the abstracted "line of communication" between sense perception and external reality: "it can transmit just what it is constructed to transmit and no more."[62] The story of the lady on Neptune and the preceding disagreement over what is happening Now would also have registered with readers on a banal level, in terms of everyday communication and travel.

By the time *The Nature of the Physical World* appeared in the more affordable Everyman edition of 1935, rationing would have faded from immediate memory. But advertising in magazines and newspapers still urged readers into incessant acts of consumer comparison. With its constant negotiation between falling apples and famous theorists, parish pumps and speeding nebulae, Eddington's text refuses to abandon everyday experience while embarking on its journey into cosmic abstraction. Through storytelling, characterization, and dialogue, much of which directly involves the reader or listener, Eddington reopened a possibility that operated in early newspaper coverage and in pulp fiction on Einstein themes. Relativity offers a set of metaphors for material and social concerns, which in turn help to draw out the new theory's comic and more profound resources. Such affinity with everyday life is harder to achieve in magazine exposition, where metaphors cannot be stabilized so easily through strong overarching principles that help to shape the reader's experience.

The rapport between fictional treatment of relativity themes and Eddington's exposition deepens when we turn to consider his own experiences of intimacy in light of the lady on Neptune story. An episode recorded in the astronomer's journal provides a clue. This document is a curiosity in itself, being neither a fully private nor a public piece of writing. Spanning the years 1905 to 1914, its entries recount Eddington's holidays and time as a student, but many of them have the feel of having been written up retrospectively, offering summaries of activity rather than slices of raw experience. The ledger is also filled with lists, including a chronology of his life, examinations, degrees, scholarships, and appointments, places traveled to and cycling tours undertaken, books read and plays seen, scientific papers written, dinners attended while at Cambridge, and sporting activities engaged in (football, fives, "tennis and spasmodically bowls").[63] Very little confidential reflection arises, but one incident does offer a glimpse into the astronomer's personal life. During late August 1908 the Eddington family spent a fortnight at Brundall on the Norfolk Broads, a customary summer visit among relatives and Quaker friends. A family named Yates, relations of the Eddingtons, were also present, and the twenty-six year old astronomer, who was by now working as a chief assistant at Greenwich Observatory, reported in his journal that a tension had arisen. "Somewhat strained relations existed between Mrs Yates and Emmeline on the one hand and my mother, sister and self on the other," he wrote, "owing to my not having done what, apparently, they had set their minds on my doing." Although Eddington is not explicit, the Yateses had clearly been thinking along matrimonial lines. Eddington feels no remorse, remarking that "My mother + sister certainly must have suffered for it more than I did, for as Rex [Emmeline's brother] ignored the tension, and my conscience did not reproach me, I got along very comfortably." The two young men "got up almost every morning at 6.15 and bathed in the river before breakfast," their friendship unperturbed by Emmeline's dashed hopes.

Eddington never married, and this diary entry suggests that his family did not put him under any pressure in this regard, even if others may have done. It is clear that he enjoyed male company, particularly if it involved shared physical exertion, and it is widely acknowledged that he had a long and close friendship with C. J. A. Trimble, whom he had met when they were both students at Cambridge.[64] Correspondence between

139

W

Mrs Henry Wood
The Channings (1911)
East Lynne (1912)

Stanley Weyman
M Red Cockade
Ovington's Bank (1923)

Williamsons
Botor Chaperon (1907)
My Friend the Chauffeur (1908)
The Princess Passes (1908)
Car of Destiny (1908)
Lady Betty (1908)
Lightning Conductor (1910)
Set in Silver (1910)
Motor Maid (1910)
Rosemary in Search (1910)
Dummy Hand (1921)

Mrs Humphrey Ward
M Robert Elsmere

Wallace
Malay Archipelago (1896)

Ward
Naturalism & Agnosticism (1906)

I Walton
Compleat Angler (1907)

Whitelaw
Pirate's Gold (1920)

P. G. Wodehouse
Indiscretions of Archie (1921)
Man with two Left feet (1923)
Sam the Sudden (1927)
Heart of a Goof (1927)
Clicking of Cuthbert (1927)
Piccadilly Jim (1927)
Adventures of Sally (1928)
Money for Nothing (1928)
Small Bachelor (1929)
Carry on Jeeves (1930)
Gentleman of Leisure
Something Fresh
Girl on the Boat (1931)
Damsel in Distress (1931)
Mr Mulliner Speaking
Big Money (1931)
Little Nugget (1932)
Bill the Conqueror (1932)
The Man Upstairs (1933)
Hot Water (1933)

H.G. Wells
War of the Worlds
Invisible Man
Time Machine
Sea Lady (1905)
Kipps (1906)
Country of the Blind (1911)
Tono Bungay (1914)
Bealby (1916)
Mr Britling (1917)
Soul of a Bishop (1918)
War in the Air (1927)
Research Magnificent (1927)
New Machiavelli (1929)
Making of the World (1931)
First Men in the Moon
Island of Dr. Moreau (1934)
Food of the Gods (1934)

H. Walpole
Prelude to Adventure (1916)
Fortitude (1916)
Jeremy (1919)
Jeremy & Hamlet (1923)
Jeremy at Crale (1927)
Perrin & Traill (1925)
Cathedral (1926)
Wooden Horse (1928)
Captives (1930)
Rogue Herries (1938)

O. Wilde
De Profundis (1905)
Lord Saville's Crime (1914)
Dorian Gray (1914)

J. M. Whitham
The Heretic (1921)

E. Wingfield-Stratford
Life (1924)

V. E. Whitechurch
Templeton Case (1924)
Dean and Jecinora (1926)
Locum Tenens (1927)
Crime at Diana's Pool (1927)
Shot on the Downs (1928)
Mixed Relations (1929)
First and Last (1930)
Murder at the Pageant (1932)
Murder at the College (1933)

Godfrey Winn
Dreams Fade (1928)

X, Y, Z

F. Brett Young
Young Physician (1920)
Murder at Fleet (1928)
My Brother Jonathan (1938)

C.M. Yonge
The Daisy Chain (1928)

E. H. Young
William (1936)

Zola
La Verité (1907)
Faute de l'Abbé Mouret (1911)
L'Argent (1914)
Germinal (1936

12. Eddington's journal. Add. Ms b. 48, p. 139. Reproduced by kind permission of The Master and Fellows of Trinity College Cambridge.

the two men has not survived, a loss that historians of astronomy regret at least as much for any insights into the development of Eddington's thinking about stellar structure as for any reflection of intimacy between the two men. A glimpse of the two men together on one of their many long hiking trips is afforded in an early biography of Eddington by his former student Alice Vibert Douglas, who quotes from Trimble's recollections of their time together. The astronomer was fond of glissading down steep hillsides, and on one occasion the descent caused serious damage to his trouser seat. "When we reached our destination we borrowed a needle and thread and though I tried to persuade Eddington to doff his garment and mend it himself he prevailed on me to do the job *in situ*," Trimble recalled.[65] "This was not too easy," he continued, "but I managed it somehow without injury to his person, though I believe I sewed him to his shirt." It is a touching image, poised between homoeroticism and close companionship.

The eclipse letters of 1919 provide another fleeting postcard of shared physical and mental pleasures in a form that likely featured intermittently throughout Eddington's life. The astronomer was clearly delighted when he encountered a young man with whom he could pass the last ten days of waiting at Madeira for passage to Principe. The fifty-year-old Cottingham had declined Eddington's invitation to participate in vigorous hiking expeditions, and did not play chess.[66] Eddington related to Winifred his good fortune in making the acquaintance of Geoffrey Turner, "a very jolly boy keen on butterflies, on swimming and on chess."[67] Aged sixteen, Geoffrey had been sent to Madeira alone for six weeks to recover from an illness.[68] The youngster, from Mumbles on the South Wales coast, showed the astronomers the best place for swimming, "a more or less enclosed pool where one can get a good swim without being knocked about on the rocks by the waves," and Eddington "got tremendously sunburnt." One evening after dinner he and Geoffrey ventured out to a picture-palace where the main feature was, rather oddly, the funeral of Edward VII (from 1910). The time Eddington spent with the youngster was undoubtedly as sexually innocent as it was brief, and we need only note his keen joy at finding an active young man with whom to pass the time. While he was careful to keep ladies with matrimonial interest at a safe distance, there is no evidence—and perhaps no need—for us to categorize his sexuality.[69] It is clear that Eddington's friendships and acquaintances, whatever physical or social form they may have taken, sustained him in a convivial bachelor existence. In the Emmeline Yates episode, it is the women of both families who bear the burden of a young man's preference for male companionship, a burden that Eddington notes without remorse. His lady on Neptune story, penned twenty years later, registers the same asymmetry. Either party to the "melancholy separation" might have been figured as the one to put in a full day's work of waiting, but for once the analogy is slanted toward the terrestrial frame of reference, leaving the only female character in *The Nature of the Physical World* to measure out the hours of separation.

The significance of this episode for readers coming to terms with shifting relations between men and women after the war becomes clearer when Lord Peter Wimsey's Eddington moment is read alongside Sayers's interlacing of popular physics with bachelorhood and male combatant experience in her detective fiction. Before moving on to a close reading of Wimsey's journey at the speed of light, there is one more term from Chief Inspector Parker's riposte to consider: "coefficients of spherical curvature." This term crops up later in *The Nature of the Physical World*, where Eddington is working toward the accommodation of spiritual experience alongside physical law. It was here that Stebbing found most to criticize, and in her book she singled out one particularly outrageous piece of anthropomorphism from Eddington's attempt to explain Einstein's law of gravitation.

Coefficients of Spherical Curvature

Having established that gravitation is now to be conceived as curvature of space and time, replacing the old Newtonian tugging force, Eddington moves on to consider the formulation of Einstein's law "governing and limiting the possible curvature of space-time."[70] Readers are primed for difficulty by the assertion that a "four-dimensional manifold is amazingly ingenious in discovering new kinds of contortion," which means that "twenty distinct measures are required at each point to specify the particular sort and amount of twistiness there."[71] These measures are known as "coefficients of curvature." Ten of the coefficients stand out from the others, and Einstein's law of gravitation states that these "principal coefficients of curvature" are always zero in empty space. An account of Newtonian law would, Eddington suggests, have to explain why tugging exists in the world. Accordingly, an account of Einstein's law has to explain why "principal curvature" is excluded.[72] Taking a step toward this intensely technical point, Eddington offers his readers a restatement of the new law: "*The radius of spherical curvature of every three-dimensional section of the world, cut in any direction at any point of empty space, is always the same constant length* [emphasis in original]." Recognizing that even the most patient of readers would by now be at the limit of their tolerance for abstruse terminology, he carries his audience through this explanation by giving it the flavor of a detective story.

How can the radius of spherical curvature always be the same? Setting aside the question of what these terms actually represent, Eddington instills confidence in his baffled readers by encouraging them to feel surprised at the basic fact of uniformity: "This directed radius which, one would think, might so easily differ from point to point and from direction to direction, has only one standard value in the world," he declares, identifying the mystery to be solved: "Why is there this unexpected standardisation? That is what we must now inquire into."[73] The explanation involves a return to elementary questions of measurement: "When we speak of the length of the directed radius we mean its length compared with the standard metre scale," Eddington observes. "We must either convey the standard metre to the site of the length we are measuring, or we must use some device which, we are satisfied, will give the same result as if we actually moved the metre rod." Here is the root of the mystery: "When the standard metre takes up a new position or direction it measures itself against the directed radius of the world in that region, and takes up an extension which is a definite fraction of the directed radius."[74] Enlarging on this point, Eddington gives the standard meter even more human qualities than Newton's apple:

> I do not see what else it could do. We picture the rod a little bewildered in its new surroundings wondering how large it ought to be—how much of the unfamiliar territory its boundaries ought to take in. It wants to do just what it did before. Recollections of the chunk of space that it formerly filled do not help, because there is nothing of the nature of a landmark. The one thing it can recognise is a directed length belonging to the region where it finds itself; so it makes itself the same fraction of this directed length as it did before.

Bewildered and wondering, recollecting and recognizing, this measuring rod was too much for Stebbing, who felt that such a "foolish picture" would "encourage the reader to believe that he has understood a theory when he has only been entertained by an irrelevant illustration."[75] She stressed that it was "extremely important that Eddington should have striven to say as clearly as possible exactly what he meant" at this point, since the sentient rod plays a vital part in establishing "Eddington's final philosophical construction," his idealist interpretation of physical law.

Anthropomorphism on its own would not have been enough to draw readers into collusion with such an abstruse argument, however. The story is given a more satisfying shape through the detective's revelation of a conspiracy: "When we felt surprise at finding as a law of Nature that the directed radius of curvature was the same for all positions and directions, we did not realise that our unit of length had already made itself a constant fraction of the directed radius. The whole thing is a vicious circle. The law of gravitation is—a put-up job."[76] This solution of the mystery helps to bridge the two parts of Eddington's book, drawing the exposition of relativity toward an argument in which the scope of physical law is heavily restricted.[77] The symbols used by physicists are shown to form a self-referential system, described as the "cyclic method of physics" and compared to the nursery rhyme "The House That Jack Built."[78] These symbols gain significance only through their contact with consciousness. Our experience of "matter," for example, is merely a product of the human mind selecting one particular set of values from the mathematical data available, singling it out for special attention. This argument, which Eddington called "selective subjectivism" in his later exposition *The Philosophy of Physical Science* (1939), had been articulated for a philosophical audience at an early stage in his engagement with relativity, not long after the eclipse results were publicized.[79] As Gifford lecturer the astronomer had an opportunity to bring out its spiritual implications: "Not once in the dim past, but continuously by conscious mind is the miracle of the Creation wrought."[80]

Readers of *The Nature of the Physical World* are prepared for this message in the book's introduction, where Eddington immediately asserts control over the two big problems that emerged in newspaper coverage of relativity: the shadowy quality of elite science and objections in the name of common sense. He insists that it is not necessary for each symbol used in science to "represent something in common experience, or even something explicable in terms of common experience."[81] Acknowledging that the "man in the street is always making this demand for concrete explanation of the things referred to in science," the astronomer pronounces that "of necessity he must be disappointed." To encourage compliance he offers an analogy between learning about physics and learning to read. Alphabet letters have no counterpart in familiar life, and appear so "horribly abstract" to children that picture books associate them with concrete images: "*A* was an Archer who shot at a frog."[82] But learners "cannot make serious progress with word-building so long as Archers, Butchers, Captains, dance round the letters." Sooner or later the child must accept that letters are abstract. In physics, Eddington explains, a similar situation has been reached: "We have outgrown archer and apple-pie definitions of the fundamental symbols. To a request to explain what an electron really is supposed to be we can only answer, 'It is part of the A B C of physics.'"[83] By 1928, Eddington had the confidence and authority to translate almost a decade of public frustration with Einstein into a simple process of growing up as a reader of the universe.

The ABC analogy feeds directly into Eddington's idealist philosophy, articulated in one of the astronomer's most frequently cited passages:

> In the world of physics we watch a shadowgraph performance of the drama of familiar life. The shadow of my elbow rests on the shadow table as the shadow ink flows over the shadow paper. It is all symbolic, and as a symbol the physicist leaves it. Then comes the great alchemist Mind who transmutes the symbols. The sparsely spread nuclei of electric force become a tangible solid; their restless agitation becomes the warmth of summer; the octave of aethereal vibrations becomes a gorgeous rainbow. Nor does the alchemy stop here. In the transmuted world new significances arise which are scarcely to be traced in the world of symbols; so that it becomes a world of beauty and purpose—and, alas, suffering and evil.[84]

Such a vision could hardly be further from Ronald Ross's condemnation of Mr. Man-in-the-Mass for daring to grasp Prospero's wand.[85] For Ross, metrical symbols were the sacred preserve of elite, specially trained adepts. For Eddington, by contrast, the symbols on their own have no sacred value, and science on its own has no unique claim to unravel secrets of the universe. It is necessary, he argues, to "recognise a spiritual world alongside the physical world" representing an equally valid domain of knowledge: "Those who in the search for truth start from consciousness as a seat of self-knowledge with interests and responsibilities not confined to the material plane, are just as much facing the hard facts of experience as those who start from consciousness as a device for reading the indications of spectroscopes and micrometers."[86] This was the message that appealed so strongly to Moore-Brabazon, who rated *The Nature of the Physical World* "the most attractive physical work that has been written in modern times" for its refusal "to take a material view of life" combined with its consideration for what "the man in the street" can understand.[87] But Eddington is very careful in his phrasing so as not to replace one orthodoxy (scientific materialism) with another (religious faith). In the draft of his book manuscript, he had referred to "consciousness as a creative moral power."[88] The revised wording creates an impression at once more stable ("seat" for "creative") and less imposing ("self-knowledge" for "moral power"). As Beer observes, he does not want to "reinstate the absolute." The admission of plural approaches to "the hard facts of experience" leaves the door open to multiple interests, not just those of the religious mystic. Eddington's message accords with the assertion of mathematically innocent access to the new space and time in Einstein-themed pulp adventure stories of the early 1920s, and his verdict appealed in turn to authors of fiction and poetry from the late 1920s onward.[89]

Eddington's expository devices draw the reader into his idealism in ways that extend beyond the Quaker convictions that drove the author's own commitment to spiritual life, creating a more widely accessible bond with consumer-readers of the interwar years. In his treatment of the FitzGerald contraction, the most memorable feature is the activity of talking, which reveals and accentuates radical divergence in experience of the world. The message about equally valid frames of reference is carried by the dialogue not only as scientific knowledge with a moral edge (as in Russell's critique of parochialism), but also as a resolution of social and narrative tension. In the discussion of time, it is individual isolation and the failure of intimacy that stand out, giving the new Einsteinian absolutes a somber aspect. The coefficients of spherical curvature represent a step toward redressing this condition, articulating a shadowy world in which the meaning of physics is self-referential, shut off in its own realm. The uncanny world of shadows is redeemed by Eddington's romance of the alchemist Mind, furnishing a happy ending in which physical law and imagination or spiritual experience can enter a harmonious relationship. It is these narrative qualities, as much as the specific analogies, that give *The Nature of the Physical World* its affinity with middlebrow fiction as found in monthly glossies like the *Strand* or *Harper's*.

We Find the Stars Are One Big Family

For authors and critics who operated in tension with the glossy magazine market, Eddington's middlebrow credentials were distinctly problematic. In "The Mysticism of Modern Science," published in the *New Adelphi* for Summer 1929, John Middleton Murry described the shadow world of physics as an attempt "to smuggle God in by the back stairs."[90] That same issue carried a review of *The Nature of the Physical World* by Geoffrey Sainsbury, who resented the book's antimaterialism, its "banality of psycho-physical parallelism," and the author's "self-complacency."[91] A more positive review appeared in T. S. Eliot's *Criterion,* in which John Macmurray acknowledged that despite its "cautious" reception among scientists and the "polite but unimpressed" response of "technical philosophers," Eddington's book was "a personal document with a universal significance."[92] What appealed to Macmurray was precisely what had rankled the *New Adelphi* writers: a "criticism of the sentiment of social materialism." Macmurray was especially keen on the prospect of a "new idealism," which he felt "must start from the reciprocity of human knowledge and human conduct, and dethrone the speculative Absolute." Eliot himself, though much interested in classical and modern forms of idealism, was not in accord with such views. In a 1932 radio talk (published in the *Listener*), he deplored the "popular attitude of hailing modern physical science as a *support* of religion" that had been inspired by Eddington and Jeans, and he regretted that "the uncritical attitude of the public towards these writers and their books" was sometimes shared by "theological writers who ought to know better."[93] This talk was broadcast the same month that the 5s. edition of *The Nature of the Physical World* was issued, marking a point at which the "uncritical attitude" was poised to spread further.

In his study of the core values underpinning Eddington's faith and his scientific work, Matthew Stanley has explored the religious context for the astronomer's response to materialism. The General Strike of 1926 was, Stanley explains, "seen by many as a dramatic sign that Christian civilisation was in mortal danger from materialist philosophy."[94] While Eddington was working to convert his Gifford Lectures into a book manuscript, the Cambridge Friends Meeting that he attended was trying to arrive at consensus over how to respond to Marxism. The perceived threat of atheist materialism from Russia would have been much in Eddington's mind as he honed his arguments for acceptance of "interests and responsibilities not confined to the material plane." The "moral materialism" that Eddington sought to oppose with his alchemist Mind and shadow scheme of physics was, Stanley reveals, "strikingly similar to the version spoken of in the Quaker community at large" and bore "no links to actual contemporary materialists," having been "formed through secondhand conversation and late nineteenth-century mechanical philosophy."[95]

The outdated materialism against which Eddington articulated his idealism was detected immediately by Aldous Huxley, who used it to depict class conflict in his novel *Point Counter Point*. Lord Edward Tantamount has managed to evade a career in politics and is pursuing instead a passion for researches into biology, consisting largely of tadpole vivisection. His assistant Illidge, a working-class man of strong socialist convictions, retains a keen affection for his employer while deploring Lord Edward's "shameful and adulterous passion for idealist metaphysics."[96] Huxley chooses the novel's most dissolute character to comment gleefully on the necessity for a class-conscious Illidge to hang onto "obsolete scientific arguments" rooted in nineteenth-century materialism: "You can't be a true communist without being a mechanist," Spandrell explains to Lord Edward's daughter Lucy Tantamount.[97] The dissolute pair are nearing the end of another night of champagne and expensive taxi rides across London as Spandrell enlarges on the communist's moral obligation to reject modern developments in cosmology:

> You've got to believe that the only fundamental realities are space, time and mass, and that all the rest is nonsense, mere illusion and mostly bourgeois illusion at that. Poor Illidge! He's sadly worried by Einstein and Eddington. And how he hates Henri Poincaré! How furious he gets with old Mach! They're undermining his simple faith. They're telling him the laws of nature are useful conventions of strictly human manufacture and that space and time and mass themselves, the whole universe of Newton and his successors, are simply our own invention.[98]

Here, as in other Huxley novels, books and their authors' names are enmeshed in the social and political tensions that drive the plot. Huxley quickly recognized the significance of Eddington's idealism as a theme through which commentators of the 1930s would articulate conflicting ideologies as they worried about who had possession of voters' minds.

Eddington's bourgeois associations were summarized almost a decade later by W. H. Auden in his "Letter to Lord Byron" (1937), published the same year as Stebbing's critique. The "Letter" sympathizes with Byron's dislike of that "bleak old bore" Wordsworth as its twentieth-century writer complains to the deceased Romantic that everybody these days seems to be climbing mountains and engaging in healthy outdoor pursuits. In an inspired rhyme, the poet matches the astronomer's name with that of a Middlesex borough representing upmarket suburbia:

> Impartial thought will give a proper status to
> This interest in waterfalls and daisies,
> Excessive love for the non-human faces,
> That lives in hearts from Golders Green to Teddington;
> It's all bound up with Einstein, Jeans, and Eddington.
>
> It is a commonplace that's hardly worth
> A poet's while to make profound or terse,
> That now the sun does not go round the earth,
> That man's no centre of the universe;
> And working in an office makes it worse.
> The humblest is acquiring with facility
> A Universal-Complex sensibility.[99]

Auden's poem was published the year after Oswald Mosley's British Union of Fascists conducted a march through East London to mark its fourth anniversary, confronted by 100,000 anti-Fascist demonstrators drawn from communist and Jewish organizations.[100] Its jogging rhythm and neat rhymes satirize the complacent values that Auden found implicated in enthusiasm for popular science:

For now we've learnt we mustn't be so bumptious
We find the stars are one big family,
And send out invitations for a scrumptious
Simple, old-fashioned, jolly romp with tea
To any natural objects we can see.
We can't, of course, invite a Jew or Red
But birds and nebulae will do instead.

But what did hearts from Golders Green to Teddington really make of Eddington's book, if they read it at all? And did all of them, men and women, Christians and agnostics, respond in the same way? The detective fictions of Dorothy L. Sayers, with their meticulous attention to the social fabric and reading habits of the English, offer a middlebrow complement to Huxley's depiction of tension between readers. Sayers herself found Eddington's treatment of science and religion very appealing, but she subjected his interpretation of Einstein's universe to intense questioning as she populated it with murderous lovers, jealous husbands, and frustrated suburban housewives.

FIVE

A Freak Sort of Planet: Dorothy L. Sayers's Cosmic Bachelors

Einstein's relativity entered British fiction as a social phenomenon, like shorter dresses or the telephone. Its relevance as a theory of the universe was secondary to the awareness of status that was sharpened whenever curved space or the fourth dimension were mentioned. Two early examples show relativity themes being used to depict anxiety about the reading masses, incorporating the new space and time among other cultural forces that their male protagonists have to contend with. When looking at how science and technology enter fiction, it's useful to distinguish between how characters feel about a particular topic and what the author believes. This is especially the case with satirical treatments. In *Potterism* (1920) by Rose Macaulay and *Crome Yellow* (1921) by Aldous Huxley, the Einstein sensation is used to aid characterization, encouraging readers to maintain a critical distance from media science. But this act of distancing plays out somewhat differently in each case, reflecting the authors' contrasting positions in relation to the mass market. Analysis of these immediate responses to relativity highlights what was distinctive about Sayers's subsequent contributions.

The Truth and Nothing but the Truth

Rose Macaulay's *Potterism: A Tragi-Farcical Tract* is probably the first British novel to register the Einstein sensation. Macaulay's twenty-three novels are not widely read today, but they were popular among middle-class readers and gained critical acclaim in her day. Alice Crawford's *Paradise Pursued: The Novels of Rose Macaulay* (1995) brings to light the mystic iconography and quest for wholeness underlying this work, which stretches from the "ordered frame" of Edwardian fiction to novels addressing war, female identity, and the struggle for faith.[1] Operating in literary terrain between her exact contemporary P. G. Wodehouse (1881–1975) and more experimental authors such as Joyce, Woolf, and Eliot, Macaulay was an innovative and popular novelist whose deft handling of her readership may grant her further attention as scholars of modernist literature continue to reconfigure the relations between high modernist, middlebrow, and mass-market authorship. *Potterism*, as Crawford has observed, incorporates the Einstein sensation "in no heavy-handed way," and it is clear that Macaulay recognized the new cosmology's duality in relation to fiction: as a contrasting cultural commodity and as a potential model for innovative literary form.[2]

Potterism allows four different characters to relate events surrounding the accidental death of a newspaper editor just after the Great War. These four sections of first-person narrative are framed by opening and concluding sections written in the third person, using what is known as "free indirect style" to dip into the consciousness of selected characters while allowing for a degree of authorial distance and comment.[3] Macaulay's Einstein allusion crops up toward the end, in a passage that enters the mind of the novel's most alienated and aloof character, Arthur Gideon. A "lean-faced, black-eyed man," whose Old Testament name connotes tearing down altars to false gods, Gideon is founder of the Anti-Potter League.[4] This small band of Oxford University students has undertaken to combat the widespread hypocrisy, self-interest, and "feeble emotionalism" typified by Percy Potter's newspaper, the *Daily Haste*, and his wife's sentimental novels, which have titles like *Socialist Cecily*, *A Cabinet Minister's Wife*, and *Audrey Against the World*.[5] Pursuing the analogy with Beaverbrook, Macaulay has Percy Potter endowed with the title Lord Pinkerton following his newspapers' service to propaganda during the war.[6] The young Potter twins have by this time cheerfully subscribed to Gideon's cause during their studies at Oxford, and Gideon himself has recovered from a near-fatal war injury to assume joint editorship of the anti-Potterite *Weekly Fact*.[7]

In contrast to the sensationalism of the Potter press, writers at the *Fact* care "for the truth and nothing but the truth," conducting "a very thorough and scientific investigation of every subject" before taking up their line, which invariably runs counter to majority opinion. Walking home through London one evening, Gideon feels a sudden desire to escape from the city's "minds crowded together" in a "dense atmosphere" that is "impervious to the piercing, however sharp, of truth."[8] He is baffled by the "mass of stupid, muddled, huddled minds," unable to make sense of their greed, ignorance, and sentimentality. In a deft twist on Macaulay's part, it is a newspaper stand that rescues Gideon from despair, offering a glimpse of how he might relate to the mass of humanity. Perusing the "placards in big black letters," he sees the headlines "Bride's Suicide" and "Divorce of Baronet" alongside a smaller, inconspicuous notice about "Italy and the Adriatic."[9] Calculating that for every person who is concerned about Italy and the Adriatic there must be "a hundred who would care about the bride and the baronet," Gideon finds it hard to "bend his impersonal and political mind to understand."[10] Is it some misplaced sense of "romance" that interests them in such topics, perhaps, or an eagerness to hear about the misfortunes of others?[11]

Gideon's critique of humanity begins to soften when he catches sight of the headline "Light Caught Bending" (a real headline from the *Daily Mail*, discussed in chapter 1). This strikes him as "more cheerful," and though he thinks it "an idiotic way of putting a theory as to the curvature of space," he is refreshed by the journalistic expectation that people will be excited by a scientific topic. He foresees Einstein's theory being discussed "with varying degrees of intelligence, most of them low, in many a cottage, many a club, many a train," and predicts "columns about it in the Sunday papers, with little Sunday remarks to the effect that the finiteness of space did not limit the infinity of God."[12] This passage is a prescient caricature of later anxieties about the influence of Eddington and Jeans, but in contrast to the likes of L. S. Stebbing, John Middleton Murry, and T. S. Eliot, Macaulay's character finds it "rather jolly" that, besides "divorce, suicide, and murder," people are interested in "light and space, undulations and gravitation." Such topics count, in Gideon's view, as "true romance" because they are connected to "the satisfying world of hard, difficult facts, without slush and without sentiment." He elevates scientists, along with "scholars and explorers," above the rest of humanity because (like *Weekly Fact* contributors) they seek truth for its own sake and "didn't talk till they knew." People might "like their science in cheap and absurd tabloid form," fed to them by a Potter press that "exulted in scientific discoveries made easy," but this is, he concludes, "better than not exulting in them at all."

Middle-class readers of *Potterism* in the 1920s may have found much to agree with in Gideon's thinking, but the text also invites the question of how far his epiphany before the placards represents a genuine rapprochement with "huddled minds." The narrative style encourages a dual response, voicing the protagonist's inner thoughts while showing how he lumbers toward comprehension of sensational appeal. His eventual approval of "science in cheap and absurd tabloid form" ultimately serves to underline his distance from newspaper readers. Gideon's attempt to "bend" his mind toward the reading masses is described just before the *Mail* headline appears, allowing readers to hear an echo that the character himself is probably not aware of. In the terms of the new cosmology, his thoughts run on straight lines, making it difficult for him to transcend parochial circumstance. The light touch of Macaulay's satire acknowledges the reader's ability to see through headlines along with Gideon, but his naïve faith in "hard, difficult facts" also encourages dissatisfaction with dry or didactic exposition. Macaulay's use of the "Light Caught Bending" headline to aid characterization makes her readership comfortable by affirming a middle ground between the Potter press and the *Weekly Fact*.

The novel's main body offers four "dimensions" to its story, narrated by Gideon himself, by Leila Yorke (in a splendid pastiche of melodramatic fiction), and by two anti-Potterites, chemist Katherine Varick and clergyman Lawrence Juke. Prejudices on the part of Potters and anti-Potterites alike are exposed, and while the narrator's opening and concluding sections do enable the differing motives and sympathies to be drawn into a single narrative, helping to resolve the question of how Oliver Hobart fell down the stairs to his death, they leave unanswered the novel's wider puzzle of how to synthesize incompatible value judgments. Macaulay toys with a narrative structure that offers itself as a literary equivalent to Einstein's theory, suggesting that each of the characters has their own way of measuring events—their own "frame of reference."[13] But any direct parallel between human affairs and scientific abstraction is evaded by the author's refusal to stand back and give an objective overview in the third-person sections narrated by "R. M." No literary equivalent to the new absolutes discovered in four-dimensional space-time is proposed. With its satirical depiction of Gideon's elitism, Leila Yorke's gushing faith in the supernatural, Katherine Varick's scientific detachment, and Lawrence Juke's compromised socialism, *Potterism* affirms the reader's awareness that any given set of values can become absurd or limiting when taken to extremes. Its refusal to adopt a neutral, unprejudiced narrative perspective suggests that it is impossible to live without absurdity in one form or another, resulting in what Crawford describes as "a middlebrow satirical novel" that is also the first English-language "Einsteinian comedy."[14] Macaulay incorporates the Einstein sensation into her experiments with self and world, using the energy accumulated through newspaper coverage to feed her conviction that laughter is the only strategy we have left for "confronting a muddled, disintegrating world."

It Seems to Upset the Whole Starry Universe

Potterism was Macaulay's tenth published novel, and she had by this time built up a relationship with a wide reading public while exploring her own fascination with pragmatist philosophies of coherence and reality.[15] In contrast, Aldous Huxley had yet to establish himself as a novelist at the time of the Einstein sensation. This difference is reflected in the way that bending (or curvature) and newspaper science are treated in *Crome Yellow*, his first novel. Here the distancing of human affairs from modern cosmology is more abrupt, and relativity themes are not allowed to engage so deeply with the novel's narrative form. The most explicit Einstein reference appears in chapter 10, during the torments endured by aspiring poet Denis Stone at Crome Manor, the country residence of Henry and Priscilla Wimbush. Crome is based on Garsington in Oxfordshire, where Lady Ottoline Morrell collected literary intellectuals during the late 1910s and early 1920s.[16] Some critics have detected a playful distancing of relativity themes even in the novel's early chapters. David Bradshaw sees Huxley opening his novel with "a witty spoof on the advent of relativity," giving the Einstein themes expounded by J. W. N. Sullivan an "irreverent treatment." Sullivan, whose *Three Men Discuss Relativity* almost won Arnold Bennett's approval in 1927, had been writing about relativity for a general intellectual audience on the pages of the *Athenaeum*, a sixpence weekly journal that was edited by John Middleton Murry from April 1919 until it was merged with the *Nation* in 1921.[17] Sullivan contributed articles on relativity during the months before the eclipse experiment results were announced, giving readers advance knowledge of Einstein's theory before the press sensation took hold.[18] Huxley and Sullivan were both working as editors on the paper, and T. S. Eliot was closely involved as a contributor.[19] Both Bradshaw and Michael Whitworth focus their analysis of Huxley's Einstein references on the social circles around the

Athenaeum, revealing how significant Sullivan's writings were in the formation of literary responses to the new physics.

The description of Denis's train journey from London to Crome is, according to Bradshaw, "a clear satire on the Einstein craze" because it combines train travel with the lead character's resentment of having "two hours cut clean out of his life."[20] Huxley may, he suggests, be using Denis's old-fashioned view of time and space to suggest that the aspiring poet is behind the times, as when he subsequently muses on a complete misunderstanding between himself and one of the female characters in adapted Euclidean terms: "Parallel straight lines, Denis reflected, meet only at infinity."[21] Whitworth sees Huxley engaging more "tentatively with Einstein and the language of geometry" in the early novels, and he observes that Denis is "subtly modifying" Euclid in this bit of geometrical musing, which may suggest that the aspiring poet is "as much behind the times as Crome" but could also gesture toward some faint hope of his being able to connect with others—to meet them "at infinity."[22] *Crome Yellow*, on this account, "hints in its first chapter that it depicts a post-Einsteinian world" through the portrayal of its protagonist's cycle ride from Crome station to the manor. Denis turns the word "curves" over in his mind repeatedly and nearly falls off his bicycle while seeking a gesture to fit the landscape.[23] Whitworth senses that this scene "alludes to curved spacetime" through a character who cannot quite get on terms with the language of the new geometry.[24] At this early stage in his writing career, Whitworth suggests, Huxley himself cannot "find a form sufficient to accommodate his knowledge of the new physics, yet is aware of the large claims made for it, not least by Sullivan." Huxley recognizes the literary potential of relativity while hesitating to deploy it.

Whether we find Huxley confident or conflicted in his satirical intent, the new space and time is clearly associated with the production and consumption of newspapers and magazines in *Crome Yellow*. Denis's halting poetic endeavors are cast into the shade of Mr. Barbecue-Smith, a journalist who boasts of churning out 3,000 words in a couple of hours. Evening entertainments underline the poet's sense of isolation and inadequacy: while Henry Wimbush grinds the tune of "Wild, Wild Women" from the pianola, a male and female guest dance together like "a beast with two backs."[25] Denis's sulking is interrupted by another of the young ladies, who insists on engaging him in conversation about contemporary verse. He has been pretending to read the *Stock Breeder's Vade Mecum* and tells Mary that his favorite poets are "Blight, Mildew, and Smut."[26] Meanwhile, Mrs. Wimbush seeks an authoritative opinion about "this Einstein theory" from Mr. Barbecue-Smith. "It seems to upset the whole starry universe," she complains, adding that the new theory "makes me so worried about my horoscopes."[27] Priscilla has recently won four hundred pounds on the Grand National, a welcome change from losing thousands in the "Old Days" before she learned to take guidance from the stars.[28]

In his mid-twenties when Einstein made headlines, Huxley used this passing reference to emphasize the caprice of a society hostess whose salon functioned as a sort of Grand National for the literary prestige of her guests. But his satire on Garsington is directed at least as much against his own youthful pretensions as it is against hack journalists or those who use horoscopes to determine a profitable course of action. Priscilla's upset starry universe is absurd, but what can Denis, whose poems all end up in the wastepaper basket, offer to compete with Einstein in the cultural marketplace? Where Macaulay affirmed a habitable middle ground for her readership between sensational headlines and didactic truth-seeking, Huxley's satire applies equally to all forms of cultural production, leaving no safe place for readers to settle.

Huxley returned to Einstein themes in *Point Counter Point* (1928), discussed briefly in chapter 4. Conversation about Einstein was by then set to become a different kind of social phenomenon, associated with the idealist interpretations of Jeans and Eddington. This development is registered by Dorothy Sayers in her novel *The Documents in the Case* (1930), which presents a crime of passion through a bundle of letters, witness statements, and local newspaper coverage. Left to speak for themselves with no intervention or framing by a narrator, these documents are as much concerned with the tension between Victorian and modern views of marriage as they are with the question of how George Harrison met his death from muscarine poisoning at a remote cottage in Devon.[29] Sayers weaves discussion of Einstein and Eddington into the increasingly tangled lives of her characters, giving the new cosmology a role to play in the suburban confrontation between traditional and progressive values.[30] The epistolary form allows for satirical distancing to be tempered with intimacy, resulting in a nuanced exploration of what the new cosmology offered readers with different outlooks.

The Loves of the Electrons

Readers of *Documents* are introduced to the Harrison household through four letters from Margaret Harrison's paid companion, Aggie Milsom, to her sister Olive. Reports on Aggie's knitting achievements and progress in the sublimation of her repressed urges are mingled with excitement about the new upstairs tenants: an aspiring writer named Jack Munting and his friend Harwood Lathom, a painter. Munting's own letters to his fiancée, Elizabeth Drake, relay the same events from an upstairs perspective. Einstein themes play a significant part in the novel, in conjunction with chemistry and questions about the emergence or creation of organic life. The murder case hangs on a difference between synthetic and naturally occurring muscarine, as suggested by Sayers's collaborator "Robert Eustace" (pseudonym of medic and author Eustace Barton).[31] In Catherine Kenney's study of Sayers, this novel is rated highly, not least for turning the chemical premise supplied by Barton, which Kenney describes as a "rather conventional, if clever, detective idea," into "the basis of a serious novel of rich social criticism and psychological complexity."[32] By producing a novel with no narrator and no appearance from the much-loved Lord Peter Wimsey, Sayers was taking a considerable risk with her readership and income.[33] The detective plot serves as a vehicle for "extended discussions" of "sexual politics, middle-class respectability, the relationship between art and life, and the possibility for belief in an increasingly secular world."[34] Is it possible for a novel to take all of this on and still be entertaining? Einstein themes help Sayers to see the challenge through, threading the novel with lively tension over whose opinions and experiences matter.

Relativity is incorporated at an early stage, via newspaper coverage, in terms that initially bear out Arthur Gideon's prediction of "little Sunday remarks" about God and the universe in *Potterism*. Margaret Harrison has been reading about Einstein in the Sunday paper, and Aggie cites Mr. Harrison's refusal to discuss relativity with his wife as yet another example of "the calm assumption of superiority that a man puts on when he is talking to a woman."[35] His insistence that Einstein is "a charlatan who was pulling people's legs with his theories" places him firmly in the nineteenth century, a point confirmed by his verdict on "the virtues of the old-fashioned domestic woman and the perpetual chatter of the modern woman about things which were outside her province."[36] Seizing an opportunity to retaliate, Margaret Harrison resumes the topic when Munting and Lathom accept an invitation to tea. Writing to Elizabeth afterward, Munting describes how he displayed "social charm" in response to Margaret Harrison's question about his view on "this wonderful man Einstein."[37] Sayers initially furnishes him with a hackneyed newspaper joke: "I said I thought it was a delightful idea. I liked thinking that all the straight lines were really curly, and only wished I'd known about it at school, because it would have annoyed the geometry master so much." But a "little stir of triumph" on the part of his hostess alerts Munting to a deeper purpose behind her question. He then "guardedly" adds his understanding that the theory has been "generally accepted by mathematicians, though with very many reserves."

In the conversation that follows, Sayers draws the materialism versus idealism debate into her portrayal of a frustrated suburban wife flirting with two young men who represent exciting and unsettling moral standards associated with modern painting and literature. Declaring that she has "always felt so strongly that materialism is all wrong," Margaret Harrison tries to establish a bond with Munting: "I do so wish I knew what life means and what we really are. But I can't understand these things, and, you know, I should so like to, if only I had someone to explain them to me." Munting's teasing reply, that she is "*really* only made up of large lumps of space, loosely tied together with electricity," is relayed to Elizabeth. Mrs. Harrison, we learn, frowns "attractively" in response, but the modern novelist quickly tires of her enthusiasm for the view that "poetry and imagination and the beautiful things of the mind are the only true realities after all." Lathom is ensnared as the conversation turns to art, marking the start of a sordid affair that will result in George Harrison's death from poison. The decline of materialism, meanwhile, meets with a positive verdict from Perry, the local parson, who declares himself "rather grateful" to relativity for making his job easier.[38] This happy verdict is subsequently cast into doubt by Munting, whose perusal of Eddington and other sources is woven into love letters expressing hopes that he and Elizabeth can somehow avoid the petty jealousies and frustrations that he sees tearing the Harrisons' marriage apart. As the situation downstairs deteriorates, he asks his fiancée: "What security have we that we—you and I, with all our talk of freedom and frankness—shall not come to this?"[39] Sayers uses his musings to put the new cosmology's spiritual associations under pressure in the context of literary and ethical pressures on an aspiring author and prospective husband.

Margaret Harrison's use of Einstein chat to try to reclaim sexual and social significance makes her seem ridiculous and poignant at the same time, a tension that is sharpened through Sayers's choice of the epistolary form. Instead of free indirect style giving access to this character's inner life, the letters offer highly opinionated views from Munting and Aggie, leaving readers to balance amused disregard for Margaret Harrison's gushing statements with the recognition that her situation and opportunities for conversation are restricted. The presence of Elizabeth, whose replies to Munting we never see, also allows more scope for self-parody on his part than could be sustained through a third- or first-person narrative: he is exaggerating traits that his fiancée may find irritating in the hope that foreknowledge will give them a better chance together. Munting's own reading in modern cosmology appears to be more detailed and at a higher level than that of Margaret Harrison, being gleaned from books rather than newspapers. But Sayers shows him grappling with the same question as his downstairs neighbor, sharing her interest in "what life means and what we really are." He may express this in more refined terms, but his reading of popular science is similarly bound up with a struggle to establish himself, as a writer and as a husband. By placing this deeper engagement alongside Margaret Harrison's superficial and caricatured enthusiasm, Sayers depicts the different types of responses available to readers in contrasting circumstances.

Munting has a second novel doing the rounds of publishers, but to earn money he is writing up the biography of an unnamed Victorian sage. Once this work has been completed, he will be in a position to resume his own literary projects and marry Elizabeth, who is already a successful novelist. The stumbling block to making sense of his nineteenth-century subject—and hence to his pursuit of matrimony—is a "hopeless contradiction" he perceives between the previous century's scientific beliefs and its "conventional ethics."[40] He wonders how the Victorians could have believed in materialism and Providence at the same time and how they managed to expound the "survival of the fittest" while practicing "a sort of sentimental humanitarianism" that, as he sees it, has "directly led to our own special problem of the multitudinous survival of the unfittest." And he cannot understand how the nineteenth century's "consolation of feeling that this earth and its affairs were extremely large and important" coexisted with a conviction that its inhabitants "were only the mechanical outcome of a cast-iron law of evolution on a very three-by-four planet, whirling around a fifth-rate star in illimitable space."[41] Voicing a twentieth-century response, Munting blends two distinct passages from *The Nature of the Physical World*: the astronomer's verdict on the scarcity of life on other planets and his description of finite but unbounded space. Nowadays, Munting supposes, it is "more reasonable" to believe in terrestrial importance than it was in the previous century, "if Eddington and those people are right in supposing that we are rather a freak sort of planet, with quite unusual facilities for being inhabited, and that space is a sort of cosy little thing which God could fold up and put in his pocket without our ever noticing the difference."[42] The new cosmology appears to dispense with the question of importance, leaving individuals to establish their own terms with the universe: "If time and space and straightness and curliness and bigness and smallness are all relative, then we may just as well think ourselves important as not," Munting concludes. Sayers wrote to her son many years later, "The reading of Eddington enlarges and disorientates the mind."[43] Munting's love letters relish that disorientation, but his light-hearted account of personal uncertainty also suggests that contemporary science may pose a threat to individual integrity. The more popular science Munting reads, the more unanswered questions he has: "I am increasingly not clear," he confesses to Elizabeth, "whether I am a mess of oddly-assorted chemicals (chiefly salt and water), or a kind of hypertrophied fish-egg, or an enormous, all-inclusive cosmos of solar-systemically revolving atoms, each one supporting planetfuls of solemn imbeciles like myself."[44] Such concerns are, he insists, "not so remote from the problem of marriage."[45] Beneath the self-deprecating humor lie serious questions. As Elizabeth Drake and Jack Munting set out to "re-continue" life, what exactly is it that they will be perpetuating?[46]

Much like Bennett in his review of Einstein books, Munting assesses the value of contemporary science in relation to past literary achievements. But where Bennett declared himself "as excited by these brief and superficial studies as I ever was by an ode of Wordsworth's or a novel of Hardy's,"[47] Munting is less enthusiastic. Completing the biography has, he complains, entailed "reading a lot of scientific and metaphysical tripe which is of no use to anybody, and least of all to a creative writer."[48] Where Lucretius and even Tennyson found poetry in the science of their day, Sayers's novelist derives scant inspiration from "all this business of liver and gonads and the velocity of light."[49] He and Elizabeth are to be married at Easter, but Munting still feels "tossed about with every wind of doctrine,"[50] unable to form any secure convictions about the significance of life on earth:

> They say now that the universe is finite, and that there is only so much matter in it and no more. But does life obey the same rule, or can it emerge indefinitely from the lifeless? Where was it, when the world was only a dusty chaos of whirling gas and cinders? What started it? What gave it the thrust, the bias, to roll so ceaselessly and so eccentrically? To look forward is easy—the final inertia, when the last atom of energy has been shaken out of the disintegrating atom—when the clocks stand still and time's arrow has neither point nor shaft—but the beginning![51]

It is only when he thinks of Elizabeth and the prospect of their life together that Munting is rescued from this overindulgence in popular science: "Our own immediate affairs are as important as the loves of the electrons in this universe of infinitesimal immensities," he affirms. This resolution follows the anthropomorphizing spirit seen in Eddington's lady on Neptune and bewildered measuring rod, carrying a recognition of the need for abstraction to be tempered by alternative interests and approaches to experience.

The unexpected success of Munting's novel *I to Hercules* has made Munting the object of journalistic excitement, and he tells Elizabeth how "the *Wail* and the *Blues* and the *Depress* and all the Sunday Bloods" are "yapping over the 'phone" to ask for his views on pressing topics: " 'Is Monogamy Doomed?'—'Can Women tell the Truth?'—'Should Wives Produce Books or Babies?'—'What is Wrong with the Modern Aunt?'—and 'Glands or God—Which?' "[52] The title of his novel and its implied contents are suggestive of James Joyce's *Ulysses*, published in 1922. But the bleak tone of Munting's musings on dusty chaos echoes T. S. Eliot's poetry, particularly the early poem "Gerontion," which concludes with the prospect of dissolving identity as individuals are "whirled / Beyond the circuit of the shuddering Bear / In fractured atoms."[53] Eliot himself had been attracted to Einstein themes at an early stage, referring in a 1919 essay on Ben Jonson to the "non-Euclidean" worlds created by artists who offer a "new point of view from which to inspect" the actual world.[54] But as Michael Whitworth has observed, this comparison was cut from subsequent printings of the essay, perhaps because such language had become too closely associated with popular talk of the fourth dimension.[55]

Munting's discussions with Perry, the local parson, are suggestive of a further connection with T. S. Eliot. When Margaret Harrison's flirtatious Einstein exchange deteriorates into suburban remarks about "great pictures," Munting turns to Perry for assistance in resolving the puzzle of Victorian materialism.[56] In a bantering spirit he also asks what Perry thinks of relativity and discovers that the parson is "rather grateful to it . . . it makes my job easier." Perry elaborates some days later, explaining to Munting that he is "thankful to find that scientists would at last allow him to believe what the Church taught."[57] Created two years before T. S. Eliot's radio talk on religion and science, Perry displays precisely those views that Eliot was coming to regret in "theological writers who ought to know better" when they looked to science for a support of religion.[58] Sayers's parson has a mathematics degree and is well versed in the writings of Eddington, Jeans, and Japp (a Catholic chemist who argued that asymmetry in molecules was evidence of a directive vital force involved with organic creation).[59] These credentials encourage Munting to take him seriously, setting aside his initial suspicion that the parson may be peddling "the usual shifty ecclesiastical clap-trap."[60] But the aspiring novelist in *Documents* ultimately disagrees with the parson's conviction that Catholic doctrines can be vindicated by modern science, dismissing the notion that "inaccurate theological metaphors can be interpreted as pseudo-scientific formulae."[61]

Sayers's depiction of Munting as a reflective consumer of popular science puts a different angle on the anxieties expressed from the late 1920s onward about the deleterious effects of reading Eddington and other popularizers who touched on idealism and mysticism. The resemblances between Munting and various prominent literary figures of the day locate him within a struggle to establish significance for modern literature in an age when glands, atoms, and whirling clouds of gas appear to exercise a tenacious grip on the reading public, overshadowing new literary discoveries. His first name Jack, his relationship with Elizabeth, and his novelistic aspirations suggest a further parallel with John Middleton Murry, also known as "Jack" and better known for his relationship with the successful author Katherine Mansfield than for his own attempts at fiction. By translating Munting to lodgings in Bayswater, however, Sayers denies him the sanctuary among like-minded intellectuals that he might have found in Murry's Bloomsbury circle. Unlike Virginia Woolf, T. S. Eliot, Aldous Huxley, or Murry himself, Munting has no circle of literary, scientific, and philosophical writers to sustain him. Instead of enjoying a chatty dinner with Bertrand Russell or J. W. N. Sullivan, he is reduced to raiding the local parson's collection of popular science books.[62] Among the likes of Margaret Harrison and the Reverend Perry, popular science themes cannot be preserved from entanglement with suburban realities.

The Documents in the Case depicts individuals attempting to pick their way through a world structured by multiple forms of writing and conversation. Munting's discussions with Perry and his letters to Elizabeth weave language and concepts from *The Nature of the Physical World* into a detailed social fabric that includes university specialists and amateurs, High Anglicans and agnostics, a hysterical spinster and a hard-working fiancé. Eddington's terms and arguments are distorted, played with, parroted, and pastiched, demonstrating that there is more than one way to take this antimaterialist text and that readers with varying levels of prior philosophical or scientific knowledge will incorporate its messages according to their own particular outlook. This appreciation of diverse reception remained important in the development of Sayers's literary career, and she gave it dramatic form in the prologue to a lecture on Dante at the Royal Institution in 1951. "Dante's Cosmos" opens with an imagined conversation between Professor Eddington and the medieval poet, who lies dying from malaria. The author of *The Nature of the Physi-*

cal World had died in 1944 at the age of sixty-two, and Sayers invites her audience to imagine him "released from the limitations of the space-time continuum" and "hastened, at a speed exceeding that of light, into the fourteenth century."[63] Comedy ensues as a feverish Dante attempts to come to grips with the world of twentieth-century physics, and it takes an entire page of dialogue for Eddington to cover three short sentences from the famous introduction to his best-selling book.

Despite his illness, the poet cannot help interrupting the astronomer's account of how modern physics treats that commonplace philosophical object, the table. Dante is an extension of the baffled reader who plays such a vital role in Eddington's expository writing, but Sayers reverses that dynamic by putting the professor's explanations at the mercy of a curious and opinionated interlocutor whose reasoning is beyond the expositor's control. Eddington is explaining that table no. 2 (the scientific version) supports his writing paper as satisfactorily as table no. 1 (the object of everyday experience), "for when I lay the paper on it the little electric particles keep on hitting the underside, so that the paper is maintained in shuttlecock fashion at a nearly steady level."[64] Dante's responses interfuse abstruse philosophy with a wider world of experience:

Dante: Particles! then we certainly have to do with sensibles. And a shuttlecock—

Eddington: Is a kind of projectile made of feathers and cork, with which people play at racquets.

Dante: A thing like a hawk's lure. Are you fond of hawking, Professor?

Eddington: That sport has gone out, I am afraid . . . "If I lean upon this table I shall not go through—"

Dante: One moment. Are you not illegitimately confusing two different orders of things—your phenomenal paper and phenomenal body with your scientific table?

Eddington: I am coming to that; "or, to be strictly accurate, the chance of my scientific elbow going through my scientific table is so excessively small that it can be neglected in practical life."

Dante: You mean that your paper and elbow are also composed of the sparsely distributed electric particles whose nature is still to be disclosed?

Eddington: That is the whole point. Does that surprise you?

Dante: Of course not. That was to be expected . . . Your audience is familiar with electric particles?

Eddington: They all know how to use electricity in practical affairs, though they have not all a thorough understanding of its scientific composition.

Dante: They know it in its effects, though not in its nature. That is likely enough . . . But I interrupt you too often.

Eddington: Not at all. "There is nothing *substantial* about my second table. It is nearly all empty space—"

Dante: How would that prevent it from being a "substance" as defined. Or by "substantial" do you mean "solid"?

Eddington: No—no—you will see presently . . . "space pervaded, it is true, by fields of force, but these are consigned to the category of 'influences' not 'things.'"

Dante: "Influences?" That is not an Aristotelian category—it is an astrological phrase. Are you an astrologer?

Eddington: God forbid.

Dante: He has forbidden it.[65]

The notion of God having debarred the twentieth-century astronomer from a career in astrology captures perfectly the gulf between medieval and modern investigation of the physical world. The leap from hawking in fourteenth-century Italy to a game of racquets in early-twentieth-century Cambridge has a similar effect, exploiting the tendency for simple physical objects to evoke social and institutional networks (today's readers may have plastic shuttlecocks and municipal sports centers in the back of their minds). The comedy of Sayers's dialogue between Dante and Eddington emphasizes the background of assumptions that can separate readers from one another, or from the literature of another age.

In the lecture that follows, Sayers pursues the theme of cultural incommensurability. "If I think my neighbour ignorant because he has not read the *Divine Comedy*," she muses, "I ought to remember that he thinks me abysmally and contemptibly ignorant for not knowing the name of a single international football player."[66] This does not mean that she intends to adjust her outlook: Sayers still regards the *Divine Comedy* as ultimately more worthwhile than international football. But she relishes the prospect of neighboring, divergent value judgments. Commenting that modern philosophers are worse off than those of Dante's time because "they are surrounded by an enormous public which is literate only in the sense that it can read print," Sayers observes that the "average modern man" has not been trained "either to understand the grammatical structure of language, or to express his meaning with precision, or to detect fallacies of language."[67] She clearly believes this to be an impoverishment, but at the same time there is evident glee in her description of what can happen to terms once they have escaped from "the hands of specialists."[68] Release a word into common use, she muses, and "it at once begins to collect associative meanings as a caddis-worm collects sticks and shells. Once the journalist, the popular novelist and the cocktail-party chatterer have got hold of it, it becomes so damaged by downright mishandling as to be perilous in any argument." The caddis-worm habits of language were not confined to the popular press, and Sayers found it "fascinating to see how even with an educated writer like Eddington, who has actually begun by defining 'substance' more or less correctly, a sort of aura or flavour of 'solidity' still clings about the word."[69]

This shading between popular usage and high culture fascinates Sayers, lending her detective fictions their sharp humor. In *Documents* there is a sense of continuity between newspaper headlines about glands, Margaret Harrison's gushing enthusiasm for idealism, and John Munting's worries about how to be a novelist, husband, and father on Eddington's "freak sort of planet." His reflections are depicted as a more subtle version of what is happening in the popular press and in conversations inspired by the Sunday newspaper. Like Eddington's use of the word "substance," however, this does not take place in a rarified realm sealed off from common preoccupations or usage. Munting's need to create a place for "immediate affairs" alongside "the loves of the electrons" unfolds in proximity to Margaret Harrison's consumption of newspaper science, just as Sayers reads Dante next door to fans of international football. Sayers's commitment to portraying a diversity of reception for Einstein and Eddington brings *Documents* closer to the literary form inspired by relativity that Macaulay perceived but chose not to realize in *Potterism*. The epistolary suburban novel shows what happens to characters who take seriously what Beer has subsequently detected in Eddington's writing: "a purposive, hard-working attempt by a physicist to establish an epistemology that will not reinstate the absolute, the out-there, in its language even as it attempts to destabilize them in its argument"[70] Sayers invites her readers to compare the experiences of male and female, traditional and avant-garde, religious and skeptical consumers when faced with popular science that "enlarges and disorientates the mind." In doing so she offers a journey equivalent to Eddington's reconfiguration of space and time according to what is "Seen-Now," drawing diverse perspectives into a coherent narrative.

Margaret Harrison's interest in Einstein and her motivation for the discussion help to signal the contrasting gendered quality of Munting's anxieties about being little more than a hypertrophied fish-egg or a collection of atoms. His experience of popular science runs deeper than that of Arthur Gideon and Denis Stone, who cannot get on terms with it as a mass cultural phenomenon. Munting's state of mind is closer to the paralysis experienced by pulp fiction adventurers confronted with the new space and time. But there is no recourse to the uncanny dream conclusion here. Sayers incorporates popular science into her protagonist's questioning of his identity and position in life, a semi-comic falling apart from which he must set about rebuilding himself as a more robust future husband. The consumption of popular science books, as opposed to newspaper science, is depicted as a distinctively male activity in Sayers's detective fictions. In light of Munting's experience, Wimsey's tendency to "go all Eddington" hints at prolonged bachelorhood. This accords with the earlier romance plots, in which an interest in the new space and time drove Major Rooke back to his rooms a confirmed bachelor, left Gilbert Merriless's wife alone upstairs, and led Hector Garden to neglect his fiancée. Through the character of Lord Peter Wimsey, Sayers developed the connection between the new cosmology and male dissociation after the Great War, using the domestic lives of her characters to question the extent to which Eddington's analogies had made the new space and time fit for readers to inhabit.

The Image in the Mirror

In her study of British fiction by women writers of the 1930s and 1940s, Gill Plain sees "the ambiguous and contradictory results of war's disruption" at work in Sayers's Wimsey novels.[71] On the one hand, the Wimsey stories display a "faith in the social responsibility of the educated (rather than the strictly upper) classes" that is "channelled into a belief in the integrity of the individual." Wimsey himself, however, represents a challenge to that faith, being "a hero whose self is clearly not a unified and coherent whole, but who is instead a fragmented, unstable and bisexual subject protected by the façade of a fluttering defensive mask." This analysis is largely based on the Wimsey novels, and Plain gathers convincing evidence from Sayers's development of his character, starting as a shell-shocked officer haunted by wartime memories and culminating in his marriage to Harriet Vane. The Wimsey short stories are often regarded as incidental to this trajectory, less significant in terms of character development than the eleven novels published from 1923 to 1937. But Wimsey's closest involvement with cosmic themes occurs in two short stories, "The Image in the Mirror" (1933) and "Absolutely Elsewhere" (1934). His skirmishes with the fourth dimension and the speed of light bear out Plain's verdict, showing how the distinctively male disorientation that Sayers associated with popular science helps depict a coded bachelor existence from which Wimsey must emerge to become a husband fit for Harriet.[72]

"The Image in the Mirror" takes its theme from H. G. Wells rather than Einstein or Eddington, but the plot is related to that of "Absolutely

Elsewhere" in its depiction of male companionship articulated through discussion of space and time. In contrast to Wimsey's highbrow Eddington blather in that story or the suburban chatter about Einstein in *Documents*, this story portrays the disturbing effects of the fourth dimension in the life of a working man who finds reading difficult. The story opens Sayers's 1933 collection *Hangman's Holiday*, which represents a hiatus in the Wimsey-Vane courtship. In *Strong Poison* (1931) Wimsey first meets detective author Harriet Vane, but there and in *Have His Carcase* (1932) she declines his proposals of marriage. He has saved her from hanging by proving that she did not murder her former lover, but this rescue creates a sense of inequality and obligation that impedes romance. Finally, in *Gaudy Night* (1935), their engagement begins, followed by their marriage in the last Wimsey novel completed by Sayers, *Busman's Honeymoon* (1937). The order of publication does not exactly follow Wimsey's personal chronology, and the short stories in particular are less fixed to this overall narrative. But during 1933–1934 Wimsey fans were left uncertain as to his future, with no mention of Harriet in *Murder Must Advertise* (1933) or *The Nine Tailors* (1934). "The Image in the Mirror" and "Absolutely Elsewhere" replace courtship with adventures in space and time, dabbling in apparently supernatural powers only to establish more reasonable explanations in terms of photography and the telephone exchange. They also build on the theme of disorientation resulting from encounters with abstract scientific knowledge, explored in *Documents*. Kenney notes that in *Strong Poison* (1931), Wimsey had been "transformed from his earlier self partially by becoming more like Jack Munting."[73] In these two short stories, Sayers allowed her bachelor to indulge in cosmic speculation without the calming presence of a fiancée at the story's margin.

"The Image in the Mirror" opens with Wimsey enjoying a drink in the lounge of a hotel somewhere outside London. On returning to his seat he discovers that the book he had brought with him has been picked up by a little red-headed man. "Not what is called 'a great reader,'" Wimsey decides, watching as the man uses both hands to turn each page, breathing heavily as he gazes at the text.[74] The volume is by H. G. Wells, and the red-headed man's attention has been captured by the story about Gottfried Plattner. Setting the book down, this stranger with a "thin Cockney voice" engages Wimsey in conversation, describing Wells as "a very clever writer" and hoping that Wimsey, having most likely "been to college and all that," may be able to help him make sense of the tale.[75] "I don't rightly understand about this fourth dimension," he confesses, adding that Wells "makes it all very clear no doubt to them that know about science."[76] As Wimsey puts it, "The Plattner Story" is "the one about the schoolmaster who was blown into the fourth dimension and came back with his right and left sides reversed."[77] Robert Duckworthy, an employee of Crighton's for Admirable Advertising, has had troubling firsthand experience of "this right-and-left business" and is badly in need of help.[78] His tale is distinctly uncanny, and Wimsey's eccentric, unbalanced behavior during the investigation delays the ruling out of irrational forces. This theme continues through the next story in *Hangman's Holiday*, "The Incredible Elopement of Lord Peter Wimsey." Both stories blend the scientific and supernatural extremes of mystery writing, revealing Sayers's mastery of her craft and her ability to stretch its conventions.

Duckworthy tells Wimsey that his trouble began when he was knocked out by a bomb blast during an air raid in January 1918. The following morning he awoke to find himself "in broad daylight, with green grass all round me, and trees, and water to the side of me."[79] The Other-World discovered by Plattner in Wells's story is dark and suffused with an eerie green light, but Duckworthy's account sounds more like the Younger London seen in pulp adaptations of *The Time Machine*. Wimsey's readiness to receive a yarn along these lines sets up a moment of bathos. "Good Lord!" he exclaims. "And was it the fourth dimension, do you think?" The man's reply brings the story sharply back to everyday circumstance: "Well, no, it wasn't. It was Hyde Park, as I come to see when I had my wits about me." But there are mysteries yet to resolve. Duckworthy reports that he lost a day and two nights, he was covered in bruises, his watch had stopped, and money had vanished from his pocket. Worse still, a girl he did not know had accosted him in a tearoom, calling him "Ginger" and insisting that he had spent the night with her. A sexual innocent, Duckworthy is distressed not least by the notion that "I'd been doing wrong and not getting full value for my money either."[80] Called up to the front soon after this incident, he was injured in another blast. On regaining consciousness in the base hospital, he found himself accused of having looted a fellow invalid who had lain helpless in a shell hole. Peculiarities continued as Duckworthy settled into civilian life after the war. His engagement was broken off when the young lady in question swore she had seen him at the seaside with another girl on a weekend when he had claimed to be laid up with influenza. And now, it appears, a man answering to his description has strangled a young woman on Barnes Common. The red-headed man fears the worst. He must have been knocked into the fourth dimension in 1918, and he has come back morally as well as physically inverted. He must now hang for a murder that he has no recollection of committing, but terror of the noose is offset by relief at knowing that these evil deeds will cease once he is dead.

Fear of a depraved double touches a chord with Wimsey, who recalls having been terrified as a child that he and his nurse would one day round a corner to find "a horrid and similar pair pouncing out at us."[81] This memory is triggered by Duckworthy's description of a recurring nightmare, caused by seeing *The Student of Prague* (a 1913 horror film remade in 1926) when he was a small child. Inspired by the Faust legend, the film depicts an evil *doppelgänger* created from the protagonist's mirror image. Duckworthy's nightmare culminates in a confrontation with his grinning double, in which he discovers that "*it* was the real person and *I* was only the reflection." The disturbing dream has recently returned: "Fighting and strangling in a black, misty place. I'd tracked the devil—my other self—and got him down. I can feel my fingers on his throat now—killing myself."[82] Duckworthy's last shred of sanity is now in jeopardy following a waking encounter with his *doppelgänger* in the mirrored door of a barber's shop.

Wimsey's empathy with this unbalanced state of mind is evident when he visits the same barbershop to investigate. Disturbed by the sight of his own reflection, he wonders what kind of man is staring back at him from the glass. A similar incident had been described by Freud in his 1919 essay "The Uncanny," but Sayers does not draw attention to the psychoanalytic connotations of this theme, leaving readers to enjoy the vignette whether or not they recognize its source. In contrast to the pulp adventure stories of the early 1920s, which used Einstein themes to explore the coexistence of primitive and modern male behavior, "The Image in the Mirror" is more concerned with inversion. The word "queer" appears several times during this story, as we might expect from a tale of strange occurrences. In the conventional period sense it simply means strange or inexplicable and does not yet carry the dominant association with homosexuality that became increasingly pronounced (and subsequently politicized) during the later twentieth century. The theme of inversion, however, does give homosexuality an unspoken presence in this story. Sexual inverts were a common topic in psychology and medicine during the late nineteenth and early twentieth centuries, most prominently in *Sexual Inversion* (1897) by Havelock Ellis. The close physical and confessional relationship between the advertising employee and his monocled confidant carries a frisson of homosexuality, appearing to corroborate Gill Plain's description of Wimsey as an "unstable and bisexual subject protected by the façade of a fluttering defensive mask."[83] Duckworthy is shy at first, not wanting to impose on his listener. "It all began—" he offers, before breaking off to glance nervously around the room. Assured that they are alone, he continues:

"There's nobody to see. If you wouldn't mind, sir, putting your hand just here a minute—"

He unbuttoned his rather regrettable double-breasted waistcoat, and laid a hand on the part of his anatomy usually considered to indicate the site of the heart.

"By all means," said Wimsey, doing as he was requested.

"Do you feel anything?"

"I don't know that I do," said Wimsey. "What ought I to feel? a swelling or anything? If you mean your pulse, the wrist is a better place."

"Oh, you can feel it *there*, all right," said the little man. "Just try the other side of the chest, sir."

Wimsey obediently moved his hand across.

"I seem to detect a little flutter," he said after a pause.[84]

The man's heart is on the right-hand side of his chest, a reversal that applies to all his other organs, including the appendix, which has since been removed. "If we was private, now, I could show you the scar," he adds. There is certainly a special moment of male intimacy here, but it has at least as much to do with the historical circumstances of Duckworthy's apparent inversion as with any physical attraction between the two men. Wimsey is deeply affected when he learns of the apparent cause:

"It's unusual, certainly," said Wimsey, "but I believe such cases do occur sometimes."

"Not the way it occurred to me. It happened in an air-raid."

"In an air-raid?" said Wimsey, aghast.[85]

The bond that goes unmentioned between Wimsey and Duckworthy is not primarily homoerotic, although there are certainly hints of this. The aristocrat and the advertising man are connected by wartime horrors that have left both of them prone to anxiety and mental aberration. This unspoken recognition underlies the shared confession of terrifying doubles and their moments of disintegration before mirrored doors.

Wimsey finally unravels the sorry tale of Emily Duckworthy, pregnant with twins when their father deserted her. Robert has been raised by his aunt and uncle as their own son, while Emily has done her best with Richard. The image in the mirrored door that Robert Duckworthy encountered was his twin brother, the real-life alternate self that a harsher upbringing might have produced had his aunt and uncle not come to Emily's assistance. The two blasts experienced during the war have left Robert vulnerable and unstable, making it all the more convenient for Richard to take advantage of his twin's existence. Both themes were close to Sayers's heart: her own illegitimate son, raised by her aunt and cousin, was by now nine years old, while her husband, who had served in the Boer War and in France in the Great War, suffered from shell shock.[86] Duckworthy's supernatural explanation in terms of the fourth dimension, gleaned from science fiction, exacerbates the effects of family separation and shell shock, leaving the protagonist susceptible to belief in a terrifying phenomenon. Wimsey restores the red-headed man's integrity by finding an alternative explanation based on the science and technology of organic and mechanical reproduction. As the polymathic detective explains to Chief Inspector Parker at the end, "The kind of similar twins that result from the splitting of a single cell *may* come out as looking-glass twins. It depends on the line of fission in the original cell. You can do it artificially with tadpoles and a bit of horsehair."[87] We are once again in the realm of fish-eggs and questions about "what life means and what we really are." Robert's culpability for Richard's crimes has been reinforced thanks to a photographic error: Wimsey proves that a photograph of Richard has been accidentally reversed in the enlarging lantern during the development process, making the twins appear identical.

The bond between Wimsey and Duckworthy suggests that the most significant difference between the fourth dimension of the 1890s and that of the 1920s was not a question of spatial versus temporal qualities, mathematical and geometrical details, or mystical power. The intervening experience of war in Europe had given male conversation about dimensional dislocation an intimate and unspoken charge that was similar—but not identical—to that of a homosexual encounter. Peter Wimsey and Robert Duckworthy have had very different family backgrounds, educational opportunities, professional lives, and roles during the war. But the conflict and its associated social dislocations have left each of them vulnerable to being lost in their own minds, unable to get close to anybody who has not shared their horrifying experience. In Sayers's writings, male involvement with disrupted time and space, across boundaries of class and education, becomes a sign of limitation on what can be shared with women. This connotation underlies Wimsey's finicking about suspects who are "absolutely elsewhere" and Parker's appeal to his friend not to "go all Eddington."

On the Back of the Light Waves

In "Absolutely Elsewhere," Sayers uses differing levels of familiarity with *The Nature of the Physical World* to signal each character's status in the

British class system. Whereas Wimsey gives an in-depth précis of a particular passage from the chapter on time, Parker's retort brings together a term used in that section (FitzGerald contraction) with a later reference drawn from the explanation of gravitation (spherical curvature). Readers who happened to have Eddington's text fresh in their minds might well have appreciated this distinction, but the contrast of tone between Wimsey and Parker would have conveyed the same broad message to those who had allowed the finer points of exposition to fade from their minds. The murder investigation centers on the service of William Grimbold's last meal, which was interrupted first by a telephone call, then by a ring at the door, and finally by the moneylender being brutally stabbed. Wimsey and Parker begin by interviewing Hamworthy, the butler, whose "spherical curvature," the reader learns directly from the narrator, "was certainly worthy of consideration."[88] Sayers addresses this joke at the butler's expense directly to her readers, acknowledging that they may well have attempted to read Eddington, in contrast to the butler, who probably has not. Like Wimsey's posh Piccadilly flat and his tendency to drop aitches, his particularity about expressions of time and space affirms a high-class status. The rotund butler, innocent of spherical curvature, represents a different kind of social "elsewhere"—a backstairs existence marking the other frontier of the world inhabited by Sayers's middle-class readers. Parker, who occupies intermediate ground between domestic service and aristocratic recreation, reflects the status of readers encountering this story in the *Strand Magazine* in 1934.

By this time Eddington's name had acquired a peculiarly double-edged quality. To those critical of "highbrowism," the Cambridge astronomer was a symptom of overly intellectual content being allowed to dominate the public sphere.[89] On the other hand, as we have seen, the brisk sales of Eddington's books were taken as a worrying sign that "common readers" were getting overenthusiastic about the new physics. The distinctively British aspect of this paradox becomes clear when the *Strand* version of the story is compared with the version that appeared a month earlier in the American magazine *Mystery*, under the title "Impossible Alibi." The technical terms from Eddington's exposition were retained, but his name was dropped, leaving Parker to expostulate, "For Heaven's sake, be reasonable."[90] The inclusion of Eddington's name in the British version of Parker's riposte functions much like the *Daily Haste* and *Weekly Fact* in *Potterism*, affirming, through Parker, a halfway position between adept use of terms like "FitzGerald contraction" and complete innocence of the new physics. The positioning of Wimsey, Parker, and Hamworthy in relation to "spherical curvature" accords with Nicola Humble's definition of middlebrow writing, which offers "narrative excitement without guilt, and intellectual stimulation without undue effort."[91] But whereas middlebrow writing tends to scorn the extremes at either side, "fastidiously holding its skirts away from lowbrow contamination, and gleefully mocking highbrow intellectual pretensions," Sayers's writing displays a teasing affection for low and high forms and their readers, gaining narrative energy from unexpected interchange between the three levels.[92]

The murder case and its resolution establish a contrast between Eddington's vast reaches of space and time and the minute details of terrestrial communication and travel. The reader's expectation that coefficients of spherical curvature and FitzGerald contractions will have nothing to do with the crime sustains curiosity through the narrative: if light-speed travel is not within the perpetrator's capabilities, what other methods were at his or her disposal? Questioned by Wimsey and Parker, Hamworthy gives an elaborate account of his master's last, unfinished meal: soup, followed by turbot served with Chablis, and then pheasant. Grimbold never receives his claret, for the butler has been interrupted by a telephone call from the moneylender's two nephews in London. Hamworthy takes the call—as the butler stresses in his interview, "it is not the cook's business, of course, to answer the telephone," and readers may imagine this spherical man hastening to meet multiple demands, his traditional service role compounded by the pressures of modern communication technology.[93] First Neville Grimbold explains that his brother wants to speak with Hamworthy. Harcourt Grimbold duly comes on the line and announces that he will be driving up to his uncle's residence, The Lilacs, later that same evening. During the call, the grandfather clock in the nephews' Jermyn Street flat is heard striking eight. Neville then comes on the line once more to ask for various items of country clothing to be cleaned and sent to him in preparation for a journey to Scotland. A visitor rings twice at the front door of William Grimbold's residence, but Hamworthy is saved from interrupting Neville's detailed wardrobe requirements when he hears the cook going to answer the door. Three-minute intervals within the telephone conversation are marked by the Exchange, and the call is renewed twice, making it (as Wimsey observes) "an expensive little item."[94]

Through the telephone system, urban places and voices are given simultaneous presence at Grimbold's country home, and each location is associated with a particular method of measuring time: the Exchange with its three-minute intervals and the London flat with its grandfather clock striking eight o'clock. A clock at The Lilacs is heard striking immediately afterward, giving the murder case a disrupted Einsteinian flavor with the suggestion of imperfectly synchronized clocks. At the end of the story, mimicking the murderous nephews' trick with the aid of his manservant Bunter, Wimsey demonstrates just how easy it is to sound as though one is in London while simply lifting the receiver of another telephone in the country residence. Having instructed Parker to gather the household in the dining room at The Lilacs, he telephones from London at the appointed hour before astonishing everybody by walking in through the library door. He has apparently completed a seventy-five minute journey in just a few seconds. Parker demands to know how he has got there: " 'On the back of the light waves,' said Wimsey, smoothing back his hair. 'I have travelled eighty miles to be with you, at 186,000 miles a second.' "[95] Following the example of Harcourt Grimbold, Wimsey has been speaking into the telephone in the library at The Lilacs, intervening in a call put through by Bunter from his own flat in London. Each call was timed so that a clock at the London residence would strike the hour as "proof" of the callers being miles away in town. The detective explains at the end that Harcourt's voice in the library really "*was* coming from London, because, as the 'phones are connected in parallel, it could only come by way of the Exchange."[96] Wimsey's apparent ability to travel at the speed of light is a marker of his intelligence and modernity, as the smoothing back of his hair nicely signals by drawing attention to his brain and sense of style.

Meticulous attention to clocks, telephones, and other markers of location in space and time is only to be expected in a detective story. But the experience of hearing clocks many miles apart chiming the hour at slightly different times down a telephone line bears a close relation to the industries of communication and time distribution in which Einstein's theory was enmeshed from its inception during the early 1900s. It is well known that the 1905 special theory of relativity was developed in the context of Einstein's work as a clerk at the patent office in Bern, where he evaluated numerous inventions for the synchronization of distant clocks by electricity, wireless, or the telephone.[97] Distribution of simultaneity was a pressing practical and business requirement, but it was also symbolic of an organized imperial power that could send electric time signals out over long distances from national observatories.[98] Consumers of relativity exposition during the 1920s would have found the discussion of time relevant to the increasing presence of clocks and time signals in their own lives. It was a popular topic: the BBC radio "pips" (or Greenwich Time Signal) were first broadcast in 1924, and *Conquest, Discovery*, and *Armchair Science* all carried substantial articles about Greenwich time and the coordination of clocks.[99] "Absolutely Elsewhere" invokes this world of distributed simultaneity, placing Wimsey's intense scrutiny of the material world in tension with his apparently distracted penchant for abstract mathematical terms. As in "The Image in the Mirror," his departure into the supernatural properties of four-dimensional space-time is a vital stage in a performance that ultimately brings readers back into the familiar everyday world. Wimsey needs to "go all Eddington" at The Lilacs, indulging in the prospect of capacities beyond the usual human range, in order to restore order in the world of butlers, cooks, and telephones.

It is more difficult to restore normality at the end of a modern detective story than after an uncanny adventure, in which the dream conclusion covers up anything too disturbing. No amount of cleverness with telephone exchanges or chat about space and time can bring the deceased man back or prevent the outcome that always recapitulates Wimsey's wartime horrors: sending a man (in this case, two brothers) to the gallows.[100] The choice of "absolute elsewhere" as a key motif gives Eddington's lady on Neptune a haunting presence in the story's background, creating an added sense of loss and loneliness for readers who have read Eddington. Aside from the unnamed cook, the only female character is Mrs. Lucy Winter, who spends the narrative's time frame traveling back from Paris and does not arrive until after the story has ended. She is Grimbold's mistress, and their union had been thwarted by her marriage to a vicious alcoholic husband who has recently died. Wimsey deduces that the murder has been inspired by the prospect of a change in Grimbold's will. The lady returning from Paris (not as distant as Neptune, but far enough away from Wapley to cause heartache), liberated at last from the shadow of an abusive marriage, ironically triggers an end to the only prospect of intimacy contained within the story. Talk about Eddington operates on a light-hearted level in "Absolutely Elsewhere" as a coded form of speech that emphasizes the close and teasing relationship between Wimsey and Parker. But Sayers crafts a narrative in which the details of Eddington's exposition are allowed to resonate more deeply with her characters' predicaments. In *Strong Poison* (1931) Parker proposes to Wimsey's sister Lady Mary, and in *Murder Must Advertise* (1933) readers learn that they are married with two small children. Parker's injunction to Wimsey not to "go all Eddington" may include a tacit plea for his brother-in-law to complete the symmetry and begin a family of his own with Harriet. By populating Einstein's universe with male readers of popular science and distant, waiting women (Elizabeth in *Documents*, Mrs. Winter in "Absolutely Elsewhere"), Sayers transforms the capacity for registering pain and disorientation latent in Eddington's expository writing into a plea for men and women to find new ways of talking to each other. By 1934, in hearts from Golders Green to Teddington, those early 1920s struggles over access to Einstein's universe had come to stand for shared male taste for cosmic puzzles, allied to the postponement of intimacy with women. Whereas pulp stories neutralized the politics of access to Einstein's universe through supernatural and romance plot devices, Sayers's detective fiction made the new cosmology problematic on different terms, substituting questions of sex and gender for those of labor interest.

In "Absolutely Elsewhere," the synchronized clocks that surrounded Einstein as he developed the special theory of relativity are redeemed from their colonization by relativity expositors and reconnected with textures of everyday life after the Great War.[101] The telephone exchange and grandfather clocks are used to question the feasibility of light-speed travel from London to Wapley, while the new Einsteinian absolutes are drawn into a depiction of fragmented relationships in postwar technological society. To achieve this connection, Sayers did not need to know about Einstein's job in the Bern patent office. She simply needed to be alive to the instability of meaning that arises when words, like caddis-worms, shift from one context to another. Sayers is, in Gillian Beer's terms, an "apt and inappropriate reader" of relativity exposition: apt in her attention to clocks, telephones, and nuances of the British class system, and inappropriate in having her detective dabble in light-speed travel.[102]

Such transposition of relativity exposition into the social and ethical terrain of fiction was possible before Eddington, as the example of May Sinclair demonstrates. Born in 1863, Sinclair was thirty years older than Sayers and Huxley and three years older than H. G. Wells. The complexity of her writing and its significance in relation to canons of modernist literature is beginning to emerge.[103] In early 1922 Sinclair offered "The Finding of the Absolute" to T. S. Eliot for his literary journal *Criterion*.[104] Though he chose not to include it, her writing is aimed at readers who are actively engaged, as she and Eliot were, with idealism and mystical philosophy. Eventually published in her 1923 collection *Uncanny Stories*, "The Finding of the Absolute" is a satire on abstract philosophy in conflict with material circumstance and affairs of the heart. It combines sophisticated pastiche of relativity exposition with challenging questions about monogamy and morality, associating the new cosmology with the revelation that traditional codes of conduct cannot deal with the complexities of human desire. The story gives its characters, who live with Kant in heaven, a cosmic vision enabling them to transcend parochialism. But, anticipating Bertrand Russell's rejection of "slavish" conduct based on physical law, this vision does not offer a workable new system of morality by which they might have conducted their affairs better on earth. The protagonist's glimpse of a fourth-dimensional morality is used by Sinclair to point to something that is missing on earth while acknowledging that cosmology alone cannot fulfill that need.

Eddington's storytelling made it fashionable to explore the gulf between cosmic and terrestrial perspectives, encouraging a broader range of writers and readers to participate in solving the distinctive early-twentieth-century conundrum of how to inhabit Einstein's universe without any guidance from the laws of physics. One young poet used Eddington even more extensively than Sayers had done, discovering in the astronomer's theories a powerful resource for coming to terms with unrequited love for a male friend. Packed with popular and esoteric reading matter and driven by rogue strategies of reading, William Empson's astronomy love poems demonstrate the ability of "apt and inappropriate" consumers of popular science to prevision new developments in science and technology.

SIX

Talking to Mars: William Empson's Astronomy Love Poems

One day in early May 1926, a mathematics student at Cambridge University walks into the Market Square and witnesses something distinctly new yet strangely familiar. The scene before him is straight out of a science fiction novel: the place is "filled with a squawking noise from a machine heard by a number of small groups of men, each of them resentfully silent in order to listen but frankly suspicious of the soft soap which was being fed to him, and exchanging as it were nudges in the pauses."[1] William Empson is due to sit his first-year mathematics examinations in three weeks' time. A fortnight later, the traditional Cambridge University season of balls and garden parties (known as "May Week," though held in early June) is scheduled to begin. But a General Strike has just been called, and the students have calculated that the festivities will coincide with the point at which supplies run out. As the nineteen-year-old Empson records in his diary, "May week is in five weeks; that will be the starvation week if we hold out till then. Will that take precedence, I wonder? Death or Dancing? These romantic pleasures!"[2] Printing has ceased, and the new medium of radio has taken on an important role: "Now there are no newspapers people get news in the marketplace or the Union, they go and stand impassively at the hoisting stations, it was done in 'The Sleeper Awakes.'"[3] This scene, Empson realized, had been described almost thirty years earlier by H. G. Wells. "There is a strong smell of that man in the air," he adds; "one prophecy, wireless, was most accurately imagined."

The Sleeper Awakes (1899) tells of a man named Graham who falls asleep in 1897 and wakes up in the year 2100 to discover himself owner of the entire world, thanks to the phenomenal interest earned on his savings. The future Britain is in thrall to Ostrog, a revolutionary dictator who keeps Graham and his nominal power safely isolated from popular unrest. Roused by the admonishments of a beautiful woman, the Victorian hero sets out in disguise to discover what life is really like for ordinary people. He comes across a hall filled with braying trumpet devices. One of them is relaying news of himself: "He says women are more beautiful than ever. Galloop! Wow! Our wonderful civilisation astonishes him beyond all measure. Beyond all measure. Galloop. He puts great trust in Boss Ostrog, absolute trust in Boss Ostrog."[4] The "Babble Machine" continues with a report of the greatly feared "negro police" who have suppressed resistance to Ostrog's regime. They "fought with great bravery," the machine announces, "singing songs written in praise of their ancestors by the poet Kipling. Once or twice they got out of hand, and tortured and captured insurgents, men and women. Moral—don't go rebelling. Haha! Galloop, Galloop!" Graham hears "a confused murmur of disapproval among the crowd" and perceives that the hall is filled with "nearly a thousand of these erections, piping, hooting, bawling and gabbling in that great space, each with its crowd of excited listeners, the majority of the men dressed in blue canvas."[5] During the 1950s Empson recalled the "exultation" he had felt on walking into a replica of this scene during the General Strike, with the curious sense of novelty and familiarity it produced around the phenomenon of wireless broadcasting: "In one way, this is quite new to me, the first time I have been in the modern world; in another way, it is as familiar as doom, because everything these men do as they listen to the machine has been described exactly in some book."[6] Wells had, he reflected, "forecast not merely the invention of the machine and its subsequent political importance but also the way Englishmen would feel about it in its political use."

These recollections come from typed drafts of an essay on the poetry of John Donne. Their purpose was to support a claim that had been made in the 1920s by F. W. Payne, the author of a small book aimed at making Donne's poetry accessible to a wider audience. At the end of his discussion, Payne had praised Donne for his "correct interpretation of the effects of the dark beginnings of our modern science," which, in his view, placed the Renaissance poet "on a level with Mr H. G. Wells for piercing insight."[7] Empson seized on this comparison, asserting Donne's science fiction credentials against prevailing values in literary criticism. "Everyone in the Eng. Lit business under forty is deliciously outraged on hearing this sentence," he wrote in 1966, "but I still think that these authors both did have piercing insight."[8] Similar remarks had been cut from the manuscript of his provocatively titled essay "Donne the Space Man" (1957).[9] "Probably the sentence about H. G. Wells is the one which excites from my young reader his highest whinny or tittering shudder," Empson speculated, "and I make bold to tell an anecdote to explain why I do not fear the same contempt."[10] His recollection of the General Strike scene followed, but was also cut from the published essay.

Empson rated Donne and Wells as science fiction authors for the predictive power of their writings. By repeatedly comparing himself and his beloved to the founders of a new world, Donne had, Empson believed, "forecast the theological quarrel about astronomy before it really began; in fact he began writing about it just about a third of a century before the trial of Galileo."[11] The published version of "Donne the Space Man" summarizes the argument in Empson's typically conversational terms: "Donne, then, from a fairly early age, was interested in getting to another planet much as the kids are nowadays; he brought the idea into practically all his best love-poems, with the sentiment which it still carries of adventurous freedom. But it meant a lot more to him than that; coming soon after Copernicus and Bruno, it meant not being a Christian—on one specific point only, that of denying the uniqueness of Jesus."[12] In his drafts Empson judged Wells's anticipation of radio and its political use more impressive: "Compared to this majestic foresight Donne cannot stand high," he conceded, "but I am trying to give the evidence for agreeing with Mr Payne that Donne did something rather like it."[13] Empson felt that Payne's interpretation needing defending against the "east wind" of T. S. Eliot, whose reading of Donne worked to exclude what Empson described as "the theology of the separate planet"—the idea that any individual may adopt the role of Jesus in a new world.[14] "One gathers that he has read Mr. Payne, and is punishing Donne for being like H. G. Wells," Empson mused, somewhat idiosyncratically. In refusing to allow Donne's poetry its preoccupation with life on other worlds, Empson felt that Eliot and his followers were suppressing politics of desire central to Donne's achievement, reducing his interplanetary concern with sexual freedom to the level of smutty jokes.

Empson was probably wise to cut this needling of Eliot from the published version of "Donne the Space Man," but his anecdote about the General Strike is worth restoring because it alerts us to a serious passion for science fiction on the part of an author who was immersed in traditional literary culture. Donne's commitment to life on other worlds was a vital and unchanging source of inspiration to Empson from his late teens until his seventies. Empson had, he explained in 1957, been imitating it "with earnest conviction" in poems written while an undergraduate at Cambridge.[15] The later essays, then, are not simply a rescuing of Donne from "neo-Christian" censorship. They defend Empson's deeply held convictions about the possibilities for science fiction that had emerged through his own poetic engagement with the new cosmology of the 1920s. Imitating Donne in his own love poems, Empson was attempting to make use of Einstein themes in the same spirit he found in the Renaissance poet's use of the Copernican revolution. His remarks on foresight in Donne and Wells are also relevant to a curious case of apparent prediction arising from the most powerful of Empson's own relativity verses. Composed between 1928 and 1935, his "Letter I" appears to describe a black hole about thirty years before these astrophysical objects were named as such.[16] Understanding how this could have happened requires a new approach to Empson's astronomy love poems, establishing their reach across the range of mass-market and elite cultural resources.

Space Is Like Earth, Rounded, a Padded Cell

One of Empson's best-known poems celebrates lovers who travel faster than light. The inspiration behind "Camping Out" is the idea, drawn from traditional love poetry, that " 'a great enough ecstasy makes the common world unreal.' "[17] First published in 1929 in the short-lived Cambridge student magazine *Experiment*, it adopts a deliberately unromantic tone:

And now she cleans her teeth into the lake:
Gives it (God's grace) for her own bounty's sake
What morning's pale and the crisp mist debars:
Its glass of the divine (that Will could break)
Restores, beyond Nature: or lets Heaven take
(Itself being dimmed) her pattern, who half awake
Milks between rocks a straddled sky of stars.[18]

Shrouded in early morning mist, the lake cannot reflect stars. It is no longer a "glass of the divine" or mirror for the heavens, as it was during the night. Instead, the flecks of toothpaste shed in the service of health and beauty create an alternative heavenly pattern, a new constellation framed by the camper's reflection as she perches on the rocks sleepily spitting into the pool. Chemicals then react with the water to produce a strange effect:

Soap tension the star pattern magnifies.
Smoothly Madonna through-assumes the skies
Whose vaults are opened to achieve the Lord.
No, it is we soaring explore galaxies;
Our bullet boat light's speed by thousands flies.
Who moves so among stars their frame unties,
See where they blur, and die, and are outsoared.

Here is the mix of theology with science fiction and everyday statement that gives Empson's early poems their characteristic element of surprise.

When Empson made a recording of his poems in 1952, he explained that "Camping Out" was "built on the one fact that if you throw soapy spots on to the surface of water they all repel each other, they move away from each other, because soap lowers the surface tension of water and the outside water pulls them, so that they look as if you were rapidly approaching a group of stars."[19] This is the reasoning of a young man who has grown up reading science fiction and popular astronomy. The poem's concluding description of blurring stars seen while traveling faster than light recalls the experience of H. G. Wells's Time Traveller as he sets out on his voyage:

The dim suggestion of the laboratory seemed presently to fall away from me, and I saw the sun hopping swiftly across the sky, leaping it every minute, and every minute marking a day. . . . Then, in the intermittent darknesses, I saw the moon spinning swiftly through her quarters from new to full, and had a faint glimpse of the circling stars. . . . The landscape was misty and vague. I was still on the hill-side upon which this house now stands, and the shoulder rose above me grey and dim. I saw trees growing and changing like puffs of vapour, now brown, now green; they grew, spread, shivered, and passed away. I saw huge buildings rise up faint and fair, and pass like dreams. The whole surface of the earth seemed changed—melting and flowing under my eyes.[20]

Empson turns the Time Traveller's sensation of the world dissolving into a metaphor for falling in love, but there is a dangerous edge to this passion in four dimensions. "Camping out together makes the lovers feel especially free from settled society, as if in a space-ship," he explained in the 1952 recording, "and on the Einstein theory this would crack up the whole of space."[21] A subsequent explanation was at once more technical and more alarming: "If any particle of matter got a speed greater than that of light it would have infinite mass and might be supposed to crumple up round itself the whole of space-time."

This explanation incorporates two distinct ideas from different books by Eddington, combining them into a single image driven by Empson's commitment to the idea that lovers can create a new world. Eddington had concluded his discussion of the "Absolute Elsewhere" in *The Nature of the Physical World* by explaining that a journey from Here-Now into the "neutral zone" was impossible because it would involve traveling faster than light: "As the speed of matter approaches the speed of light its mass increases to infinity, and therefore it is impossible to make matter travel faster than light."[22] Impossible journeys and outlawed spaces particularly appealed to the young Empson, whose love life tended to run against convention. A more extreme condition of space-time had been offered—and ruled out—in Eddington's earlier book, *Stars and Atoms* (1927). Here it was explained that stars of high mass could not be very dense, for under "the modern theory of gravitation" this would lead to strange conditions.[23] First, the astronomer noted, "light would be unable to escape; and any rays shot out would fall back again to the star by their own weight." A second consideration arose from the pattern of dark bands or lines seen when starlight is passed through a prism to reveal its spectrum. Einstein's theory of gravitation states that these lines "will be slightly displaced towards the red end of the spectrum as compared with the corresponding terrestrial lines."[24] The displacement increases in stars of high mass and small radius, and in a very massive dense star "the Einstein shift . . . would be so great that the spectrum would be shifted out of existence."[25] Eddington found these first two points tolerable, but a third consideration ruled out all possibility of such objects existing: "mass produces a curvature of space," he reminded his audience, "and in this case the curvature would be so great that space would close up round the star, leaving us outside—that is to say *nowhere*." Empson's poem links these two observations about high mass, dispatching its ecstatic lovers on a voyage into the Absolute Elsewhere that leaves the "common world" nowhere. With its impossible and extreme forms of space, Einstein's theory appeared to afford even more radical romantic escapades than the Renaissance argument for a plurality of worlds.

Being a space adventurer was not all fun, however, as a suddenly heartfelt comment in "Donne the Space Man" indicates. Empson noted that Catholicism, with its belief in saints, had made it easier for Donne to think about alternatives to Christ. But "by the time he took Anglican Orders," Empson mused, "I imagine he was thankful to get back from the interplanetary spaces, which are inherently lonely and ill-provided."[26] Something about Einstein's universe had tinged his own exploration of these spaces with a darker edge, first elaborated in "The World's End." This poem first appeared in the *Cambridge Review*, an established weekly paper of the university, in May 1928 under the title "Relativity."[27] The poem's key source is Eddington's description of "finite but unbounded" space in chapter 4 of *The Nature of the Physical World.* The expositor approached this notoriously difficult concept by comparing four-dimensional curved space to the earth as a three-dimensional sphere. On the surface of the earth, he observed, "points reached by travelling vast distances in opposite directions would be found experimentally to be close together."[28] It was the same in four dimensions: space was "boundless by re-entrant form not by great extension," and its "boundlessness" had nothing to do with its "bigness."[29] The astronomer turned to Shakespeare for a memorable image to capture this condition: "*That which is* is a shell floating in the infinitude of *that which is not* [emphasis in original]. We say with Hamlet, 'I could be bounded in a nutshell and count myself a king of infinite space.' "

Empson's poem begins with an exciting invitation to explore space, but freedom quickly yields to a realization that there can be no escape from Hamlet's nutshell. The poem opens with four lines of reported speech:

"Fly with me then to all's and the world's end
And plumb for safety down the gaps of stars
Let the last gulf or topless cliff befriend,
What tyrant there our variance debars?"

Alas, how hope for freedom, no bars bind;
Space is like earth, rounded, a padded cell;
Plumb the stars' depth, your lead bumps you behind;
Blind Satan's voice rattled the whole of Hell.[30]

In his note to the poem, Empson identified *Paradise Lost* as the source of Satan's anguished cry: "He called so loud that all the hollow deep / Of Hell resounded."[31] "The World's End" forecasts a bleak prospect for those seeking to escape the earth, comparing them to Tantalus, who was unable to drink from the river ebbing below his chin or eat the apples above his head:

Apple of knowledge and forgetful mere
From Tantalus too differential bend.
The shadow clings. The world's end is here.
This place's curvature precludes its end.[32]

Language textures from popular and technical science books are blended with ideas from the literary classics as if there is no distinction to be drawn. "The shadow clings" echoes Eddington's verdict in the lady on Neptune anecdote: "From this point of view the 'nowness' of an event is like a shadow cast by it into space, and the longer the event the farther will the umbra of the shadow extend."[33] In a typically double-edged conclusion, "precludes" offers two opposite yet equally dismal possibilities, each amounting to the fact that there can be no escape: the universe has no end, and it is already closed.[34] "The World's End" enacts the sense of promise followed by paralysis that was associated with the new time and space in pulp adventure stories, and Empson imports traditional literary characters and themes into the hazardous new universe. The poem works both ways, recasting the pain of Satan and Tantalus into a more intense modern form while using familiar literary allusions to bring out the new cosmology's starkly inhospitable aspect.

The two mythical characters are also suggestive of a parallel with the experience of audiences faced with exciting new forms of specialist knowledge. Empson's blind Satan banished from heaven and Tantalus craning after the apple of knowledge are in the position of readers who have seen the stars and the news headlines, but have been told that they do not have the mathematical tools to understand how the universe works. That devastating pun on "precludes" may apply to the changed scope of physical knowledge in the early twentieth century. Mathematics and physics had been extended through philosophy and into religion, and it was now possible for any kind of universe to be dreamed up. There was apparently no end to the scope of these speculative endeavors. And yet this realm of imaginative exploration was foreclosed to those without the requisite qualifications or membership in learned societies. "This place's curvature precludes its end" translates the physical curvature of four-dimensional space-time into a metaphor for technical knowledge that is apparently unbounded in its epistemological possibilities, yet finite in terms of qualified participants. Having grown up during the years of Einstein sensation and exposition alongside new prospects for space travel, Empson was drawn to Donne's poems as a defiant example of how nonspecialists might explore the universe on their own terms. His insistence on the capacity of lovers to create a new world is driven by rejection of the idea that physics and astronomy should be left well alone by the uninitiated. With his Cambridge University mathematics degree, Empson was hardly a mathematical innocent. But in his own poetry he adopted a form of working innocence that enabled him to apply "other knowledge, other questions" to some of the more abstruse features of Einstein's universe.

These poems celebrate readers' right to make their own way among the branches of knowledge, pursuing connections and leaps from which the focus of the specialist is debarred. But could any other reader follow the meandering path through literary classics, contemporary astronomy, and science fiction that gave rise to Empson's perplexing verses? With their dense packing of disparate material, the love poems Empson wrote while he was an undergraduate at Cambridge University risk never being understood by more than a handful of readers. Their publication in university magazines ensured an appreciative (if often baffled) audience, but their allusions and twists proved challenging even for those who were reading many of the same books Empson had rifled through.

"Camping Out" was published in the second issue of *Experiment*, a new magazine started by a group of Cambridge University students who wished to explore across the sciences, philosophy, and the arts.[35] The magazine, which ran for a total of seven issues from 1928 to 1931, was financed by Lord Ennismore (William Hare), whose father had removed him from Oxford to Magdalene College, Cambridge, in the hope that he might be cured of socialism.[36] It didn't quite work: Ennismore was active in the House of Lords as one of the few Labour peers during the 1930s, and he went on to become governor-general of Ghana from 1957 until the declaration of a republic in 1960.[37] Other members of the group included pioneering documentary filmmaker Humphrey Jennings, experimental poet Hugh Sykes Davies, and Jacob Bronowski, who went on to present *The Ascent of Man* (1973) and other popular science television programs. A rival magazine named *Venture*, launched at the same time, was more strictly literary and less interested in science, cinema, or experimental writing.[38] *Venture* included more conventional poetry, featuring work by Julian Bell (Virginia Woolf's nephew, killed in the Spanish Civil War) and actor Michael Redgrave.

A series of single-poem pamphlets published by Cambridge bookseller Heffers in 1929 featured pieces by Empson and Bronowski alongside contributions from the *Venture* stable. Reviewing the series, one Cambridge student newspaper deplored the constant posing of its prominent contributors. Poets were the worst sort of university intellectuals, the *Gownsman* reviewer complained, because they were constantly making an exhibition of themselves, in contrast to the more sober conduct of their scientific counterparts. The "brilliant thinker in Natural Science, the future Thomson or Eddington," this commentator noted with approval, "does not wear a distinctive dress," nor does he "converse publicly in such a loud voice and with such an air of arrogance that it sickens the ordinary man, and though he may have ideas on humanity as advanced and as intellectual as a prominent aesthete, does not disregard the fact that his ideas may be wrong."[39] The future scientific genius, he continued, "does not drink foul drinks with foreign names, for the sake of being drunk in Russian rather than tight on beer." The *Gownsman* reviewer found it "particularly disappointing" that "pose" was creeping into Empson's poetry, giving rise to the "suspicion that he is deliberately trying to puzzle and to shock his readers." A similar verdict was reached by Cambridge's other main student paper in June 1929, by which time the two rival literary magazines had each produced their third number. Readers would, the *Granta* reviewer warned, "find in *Experiment* a large number of lamentably clever young gentlemen talking clap-trap."[40] In *Venture*, "which will aggravate them much less, and reveal its ethical inferiority to *Experiment* by doing so," they could expect to find "few who are clever at all." *Granta* was more tolerant in its final verdict than the plain-talking *Gownsman*, concluding that there was "a great deal of real merit" to be found in both magazines.

The depiction of *Experiment* as containing deliberate puzzles or claptrap, and as being shocking or "ethical," reflects aspirations on the part of its contributors to reach beyond the usual run of Cambridge student interests, setting their sights on Paris and other hotbeds of experimental art and new morality. The more down-to-earth *Gownsman* continually mocked this ambition, primarily through an association between prominent members of the *Experiment* group and that infamous contemporary symbol of suburban propriety, the aspidistra.[41] "Does your bedder [college domestic servant] know of this?" demanded a January 1929 headline announcing the Aspidistra Advancement Association, Inc, complete with prospectus and list of directors, including "Bombardier-Grenadier Sir Bloodstock-Gore" and "His Lordship the Bishop of Six Mile Bottom."[42] On the facing page the following sonnet appeared:

Scorn not the Aspidistra. We have frowned
Mindless of vasty import. To this tree
Bedders unbare their hearts. The memoree
Of this sad shoot gave ease to Empson's wound;
From 'neath its boughs did Grigg the world astound;

With it Bronowski soothed an exile's grief.
Why th' Aspidistra reared a verdant leaf
Over the Fool's Cap which the Granta bound
On its receding brow. A noble gamp,
It cheered mild Pentland, called from Slumber-Land
To woo the Union's grace. But when the damp
Closed round the paths of Cambridge, on its stand
The Thing became a Nuisance, whence we drew
Soul-devastating thoughts, alas, too blue![43]

The *Gownsman* aspidistra motif reduces experimental writers and union debaters, bedders and brigadiers, to the same state of mediocrity. In January 1931, a full-page cartoon headed "Home Thoughts from Abroad" appeared with the caption "Cambridge poet deeply moved by sight of aspidistra in Paris."[44] The cartoon depicts a moustached and well-tailored figure clutching a pipe in one hand and his breast in the other, going into raptures as a cart is wheeled past the "Au Lapin d'Or" café with three aspidistras perched on the back. As Jason Harding has pointed out, the *Experiment* group's contribution to the Paris-based magazine *transition* in June 1930 was laced with "nervous shifting," showing "that a native strain of Cambridge rationalism could never quite be happily transplanted to the heady soil of Parisian bohemia."[45] The *Gownsman* also kept up a running commentary on developments in the facial hair of *Experiment* contributors. In March 1930 they issued congratulations to "Mr. Bronowski, for his moustache etcetera," and in May it was announced that "The Cult of the Moustache" was assuming "threatening proportions amongst our aesthetic brethren. Following on the inspiring lead of Mr. Bronowski, Mr. Davenport is now arrayed in full glory."[46]

By the time these crazes had got under way, Empson had already been dismissed from a junior fellowship at Cambridge following the discovery of contraceptives among his possessions.[47] But the *Gownsman* teasing identifies a theme that is central to his undergraduate love poems: tension between shared values and outsider status. The most intense exploration of this theme arises in a poem simply titled "Letter," printed as the second item in the first number of *Experiment*. Before its publication in the new magazine for clever young gentlemen, however, it was typed up and sent to a reader whose own studies at Cambridge would have enabled him to appreciate the poem's worrying about a common experience only too well.

Empson's five "Letter" poems are addressed to an unnamed recipient whose sex is left unspecified. John Haffenden has revealed the story of Empson's affection for Desmond Lee, but has also warned that "the discovery of the concealed identity of the beloved in these poems makes little difference to one's interpretation of them."[48] They must, "like any love poem . . . be explicated and evaluated, not by their being encoded in a personal fashion but by their generic or universalizing power."[49] The Letters translate personal despair into a universal condition by dramatizing disparate reading experiences. "Letter I" roams desperately through books, from Pascal's *Pensées* and stories about life on Mars to the theoretical intricacies of Eddington and studies of primitive religion. "Letter II" attempts to read the beloved's face as if it is a series of cave paintings, finding that the frescoes fade as light is brought in to read them. "Letter III" ranges from architectural blueprints through Gothic fantasy and advertising signs to biblical and classical mythology. "Letter IV" returns to Eddington, searching among the stars for an alternative to conventional monogamy.[50] The fifth and final "Letter" in the original sequence takes refuge in the cool abstractions of geometry.[51] Bringing Martian wireless contact and advertising into conjunction with Milton and primitive Greek religion, these poems have as much to say about reading and readers in the modern world as they do about unrequited love. Like "Camping Out" and "The World's End," they do not make for easy reading, but they maintain a sense of humor and are passionate about being baffled in ways that can be infectious.

Dark Spaces between Stars

"Letter I" opens with a famous quotation about being afraid of space. The speaker is reconstructing a conversation between himself and the person he's writing to:

> You were amused to find you too could fear
> "The eternal silence of the infinite spaces."[52]

When preparing the notes for his first collection of poetry in 1935, Empson had no need to identify the source of this line. Quoting from Pascal's *Pensées* has long been a way to signal an intellect grounded in emotional sensitivity. The frequently quoted phrase—in French, *le silence éternel de ces espaces infinis m'effraye*—epitomizes claims to humanized learning.[53] But what counts as "human" could vary considerably from one reader to the next, and the presence of Pascal in a twentieth-century conversation could signal widely diverging interests.

Aldous Huxley exploits this quality repeatedly in *Point Counter Point* (1928). While the spinster Beatrice Gilray mends a pink camisole beneath the ticking clock, Walter Burlap's annotated copy of the English translation of Pascal's work, *Thoughts*, lies open on the table.[54] Its presence invites the reader to wonder to what degree Burlap's self-regarding spiritual heat will melt Beatrice's cool detachment or ease her fear of physical intimacy. Three chapters later we learn that the French original has, alongside Carlyle, Whitman, and Browning, featured among the influences on novelist Philip Quarles. But this learning only serves to underline his perpetually transient sense of self: "Pascal had made him a Catholic—but only so long as the volume of *Pensées* was open before him."[55] In contrast, Pascal and William Blake sit unread on the bookshelves of Spandrell, who eats his way, "like a dung beetle's maggot in its native element," through every last inch of the "enormous crackling pages of the *Times*."[56] Quarles the novelist steers a middle ground between Spandrell's cynicism and Burlap's spiritual onanism. But he is incapable of genuine passion for any system of ideas, or of real intimacy. Quarles is a successor to Denis Stone from *Crome Yellow*, further advanced in his literary endeavors yet hampered by lack of conviction, a state that his reading habits only seem to exacerbate. This is precisely the condition afflicting John Munting in *The Documents in the Case*, as Sayers acknowledged by having her protagonist cite Huxley's novel. "I am tossed about with every wind of doctrine," Munting confesses to Elizabeth, "and if I'm not damn careful I shall end by writing a *Point Counterpoint*."[57] Empson's love poem takes a different approach to the problem of too much reading and not enough belief. The poem's regular pattern of rhyme and rhythm works its way through four successive attitudes toward space, each one crafted from a different type of published matter. Exploring fear, approval, doubt, and finally, extreme isolation, the poem translates an individual case of unrequited love into a universal puzzle about distance between readers.

"Letter I" opens in the same vein as "The World's End," reveling in the "eternal silence of the infinite spaces." A fixed rhyme scheme allows each line to delve into a further layer of description, digressing extravagantly across line endings that hold everything together:

That net-work without fish, that mere
Extended idleness, those pointless places
Who, being possibilized to bear faces,
Yours and the light from it, up-buoyed,
Even of the galaxies are void.[58]

Every phrase has multiple meanings, adding to the sense of excessive possibility. The "pointless," star-free places might, if they were occupied, be filled with constellations, reminders of heroic deeds that are surely preferable to sitting at a typewriter, hammering out cosmic rhymes. The digressive lines are given a focus through the speaker's obsession with his beloved's face. Following poetic tradition, he might compare it to the sun, which could be part of a constellation viewed from millions of miles away. When a lover smiles, his face lights up and such illumination could be reflected back, increased and "buoyed up" by a returning smile—just as sunlight is reflected in the moon, which then appears to shine for itself. But there can be none of that tonight. Space is empty, utterly void.

Desmond Lee had arrived in Cambridge to study classics in the autumn of 1927, just as Empson was embarking on his final year of mathematics. Perhaps encouraged by his new friend's interest in popular science, Lee quoted the line from Pascal in one of their early conversations.[59] Empson's lively disagreement may have been fueled by his own reading of popular essays by Cambridge geneticist J. B. S. Haldane, nephew of the viscount who had Einstein to dinner in 1921. At the start of *Possible Worlds*, which Empson reviewed in January 1928, Haldane gave short shrift to Pascal's "irrational terror." The "silence of interstellar space" was, he complained, a meaningless proposition: "One could not live in it, and hence could not discover whether it is silent or not."[60] Wondering how to approach his handsome friend, Empson crafted a more measured response:

I approve, myself, dark spaces between stars;
All privacy's their gift; they carry glances
Through gulfs; and as for messages (thus Mars'
Renown for wisdom their wise tact enhances,
Hanged on the thread of radio advances)
For messages, they are a wise go-between,
And say what they think common-sense has seen.[61]

Distance is good, the speaker suggests, because it leaves room for interpretation. Thanks to these dark spaces, you and I can maintain incompatible positions. You may continue to interpret my lust as friendship, and I shall continue to read a more than platonic passion into your glances. Or, if I am sensible, I will try to convince myself that your regard is purely platonic in nature. And if you have any common sense whatsoever, you will know from the way I return your gaze what my true feelings are. To show

how amused I am by the situation, I shall enact this comedy in the form of my verse. Look—I have inserted a digression about life on Mars in the middle of my argument about how communication works, but you can safely ignore that and still get the gist of my lines. I shall repeat the word "messages" to underline how easy it is for us to understand one another about this—and, of course, to fill out the carefully balanced rhythm of my verse, nicely demonstrating the state of mental equilibrium I have reached in considering our situation.

In the note to "Letter I" that Empson supplied for the 1935 edition of his *Poems*, he intensified the Martian puzzle while purporting to clarify it. "The *thread*," he said of that parenthetical comment, "was meant to be 'the unlikely chance that we *never* learn to talk to them by radio and thus find out that they are not wise [emphasis added].' "[62] The prospect of interesting planetary neighbors will probably be dissolved once radio communication discovers a lack of intelligent life on the red planet. The incomplete development of wireless technology is, according to this somewhat contrary outlook, a thread of hope from which our fantasies of contact dangle precariously. A more violent sense of "hanged" imposes itself on the poem: technological advances will strangle the clever Martians of our dreams. The poem's proposed interplanetary love affair is similarly doomed: the more clearly its recipient understands its message, the more likely he is to pour cold water on the sender's hopes for reciprocated passion.

Astronomical research into life on Mars had held a prominent place in the public imagination for over half a century, spurred by the observation of "canali," or channels, on the planet's surface by Giovanni Schiaparelli in 1877.[63] Popular discussion during the early 1920s often centered on the imminent prospect of wireless contact, thanks largely to the showmanship of Italian radio pioneer Guglielmo Marconi. "Einstein on the Mystic Wireless—What Martians Would Do," ran a *Daily Mail* headline at the end of January 1920, over a story reporting Einstein's view that neighboring planets might well be inhabited.[64] The "celebrated German astronomer" was characteristically reserved in his verdict on the mysterious signals lately received by Senator Marconi. Intelligent creatures would be more likely to use light rays than wireless signals, Einstein suggested, and he concluded that "the present state of our own technical equipment" made it "impossible to hope to satisfy ourselves of their existence."[65] Successive reports of mysterious radio signals received simultaneously on both sides of the Atlantic inspired continued speculation and invention on Martian themes, including the novel *Aelita* by Russian science fiction author Aleksei Tolstoy. The novel was not translated into

English until 1950, but a lavish 1924 film adaptation, featuring stunning constructivist set designs, was widely publicized.[66] It tells the story of a disillusioned engineer who builds a rocket to Mars, where he falls in love with its queen, who is already smitten with him thanks to stolen glimpses through advanced Martian telescopes. In August 1924, during the opposition of Mars, radio silence was observed at stations in Europe and America in case the Martians took advantage of their rare proximity (35,000,000 miles) to send a message.[67]

Life on other planets and futuristic applications of wireless were of course common themes in pulp science fiction. In the summer of 1925, Coutts Brisbane contributed a two-part adventure story to the *Yellow Magazine*, which opens with the achievement (ten years hence) of wireless telegraphy between Earth and Mars.[68] After a further interval of thirteen years, during which a common code is painstakingly established, the "Rayway" is set up. This system enables one-way physical displacement, from Earth's surface to Mars, and it is only a matter of time before the scientist Gregory Penley and his wife Florence find themselves accidentally transported. Unfortunately, a flask of Mrs. Penley's culinary invention—a revolting invalid jelly, rich in "proteids"—gives strength to the red planet's tree monsters, and the remaining intelligent beings of Mars are all slain. The scientist and his wife remain trapped on Mars, with hopes of a Rayway back to Earth fading. Empson may not have happened across this story in the *Yellow*—he was by this time eighteen years old and about to take up a place at Magdalene College to study mathematics—but Brisbane's narrative reflects a widespread contemporary interest in Martian contact. In his inimitable style, the pulp author inverted the plot of H. G. Wells's *War of the Worlds* (1898) by introducing a dubious health food supplement concocted by the strong-willed wife of an underrecognized scientist. Brisbane's satirical touches recast Martian speculation in terms of domestic dynamics and patent medicine advertisements, both familiar to readers of the *Yellow* and other popular magazines. The dismal outcome of Penley's adventure is cheery, however, in comparison to J. B. S. Haldane's dark speculation in "The Last Judgment," the final piece in *Possible Worlds*. Searching for a new abode, a terrestrial expedition reaches Mars in the year 9,723,841 but is unable to communicate with the dominant species, who are "blind to those radiations which we perceive as light, and probably unaware of the existence of other planets."[69] The expedition, and another one following it, are annihilated, thanks to Martian access to senses beyond the human range.

Empson turned fourteen in the autumn of 1920, having taken up a scholarship at Winchester College. Martian wireless aspirations would have been difficult to avoid throughout his teenage years, and Winchester students—known as "Wykehamists"—may well have circulated copies of *Discovery* and *Conquest* acquired during school vacations and leave-out days. *Discovery* included "Our Neighbour Worlds" in 1922 and "The Story of Mars" in 1924. *Conquest* featured "The Map of Mars" in 1920 and "Life on Other Worlds" in 1923. The first of these, by amateur astronomer Joseph H. Elgie, stressed the contribution amateurs had made in helping to establish details of the red planet's geography. The seas, lakes, promontories, and desert regions had, he remarked, been awarded "such weird-looking names" that one might "almost fancy that the 'Martians' themselves had wirelessed the names to the earth as being the most appropriate to their strange world that they could invent."[70] Three years later, Henry Spencer Toy contributed four pages of enthusiastic discussion in his article "Life on Other Worlds," declaring it "almost certain" that "among the distant worlds are those that teem with life."[71] The last two pages of this article were devoted to the most captivating planet in our solar system: "For a very large number of people," Spencer Toy acknowledged, "the whole of astronomy is summed up in one question, 'Is there life on Mars?' " He went on to outline the theory of the late Professor Lowell, "the leader of those who believe that Mars is the abode of an intelligent race," who had conjectured that seasonal growth of vegetation could be discerned along the canali. Four years later, Eddington acknowledged that "Martian natural history is not altogether beyond the limits of serious science," while noting that few astronomers would follow Lowell "all the way on the more picturesque side of his conclusions."[72] The Gifford lecturer did not think "that the whole purpose of the Creation has been staked on the one planet where we live" and believed that "in the long run we cannot deem ourselves the only race that has been or will be gifted with the mystery of consciousness." But he did feel "inclined to claim that *at the present time* our race is supreme" and conjectured that "not one of the profusion of stars in their myriad clusters looks down on scenes comparable to those which are passing beneath the rays of the sun."

With these debates in mind, a further layer of meaning emerges from the first verse of "Letter I." The "net-work without fish" is suggestive of canals, which in Spencer Toy's article are described as a "network" (probably inspired by Percival Lowell's striking 1905 map of Mars).[73] One buried meaning of "mere" is a marsh, pool, or sea, and "pointless places" might be conceived as planets fit for life, "possibilized to bear faces" yet uninhabited. Light from such a planet's sun might be "up-buoyed" by reflection in the planet's pools of water (ruled out by the twentieth century, but much discussed in Galileo's time). Or it might be refracted through the atmosphere. Spencer Toy described the earth's atmosphere as "like a great encircling ocean" and remarked that the atmosphere of Venus was "so dense that it is a matter of doubt if any possible inhabitants would ever see the sun."[74] A reading of the second verse in terms of life on other worlds lurks beneath its elaboration of empty space, contributing a starker sense of emptiness to the concluding line. With hopes of inhabited worlds in mind, the "eternal silence of the infinite spaces" takes on a more literal aspect, there being nobody else out there to talk with.

Alive to popular interest in the Martian question, Empson's poem uses this theme to give the speaker's passion a greater poignancy than could be attained by simply quoting from Pascal. Wireless had become a familiar medium by the late 1920s, but the poem suggests that listeners will be most receptive to messages not yet received. Just as technological advances are at their most fascinating immediately before their full potential is realized, desire may be more compelling just before the reality of a direct encounter. "Letter I" gives each of these disjunct forms of excitement a sharper taste than they could have separately, using the traditions of love poetry to give Martian dreams a tragic aspect, while popular interest in radio technology helps to keep the stellar romance from becoming overidealized. The pun on seduction in "radio advances" also suggests an affinity with the "wireless of the heart" found in pulp fiction magazines of the mid-1920s, transmuted into "love waves" by Professor A. M. Low in 1930.[75] The flexible cultural alignment of Empson's poem may be contrasted with the musings of Arthur Gideon in Rose Macaulay's *Potterism*, whose claim to an affinity with mass culture only served to underline his distance from the "mass of stupid, muddled, huddled minds."[76] The poem is, however, acutely aware of the gulfs that can spring up between readers. In the second verse's optimistic terms, such differences "say what they think common-sense has seen." But the prospect that two potential lovers may be inhabiting different planets casts doubt on the possibility of shared understanding, driving the poem away from its empathy with popular culture and into deeply esoteric fragments of scholarship local to Cambridge University in the 1920s.

No "Physics" between Them

On receiving the typed "Letter" poem from his eccentric friend, Lee might have been shocked into silence by its implication. But he could not have helped laughing at the audacity of its third verse. Here, in rhyming lines,

Bill had created a marriage between the two men's current obsessions: the primitive religion taught by Francis Cornford on the classics syllabus and the non-Euclidean space described by Arthur Eddington. Key terms from these two authors are woven together, ironically formulating an impossible division between the two students:

Only, have we space, common-sense in common,
A tribe whose life-blood is our sacrament,
Physics or metaphysics for your showman,
For my physician in this banishment?
Too non-Euclidean predicament.
Where is that darkness that gives light its place?
Or where such darkness as would hide your face?[77]

Burying his message in a web of esoteric references, mixing up analysis of primitive religion with the latest cosmological speculation, the speaker takes refuge in a personalized code derived from intensely specialized research. These lines elaborate the loss of a connecting medium, forming a desperate counterpart to that careless reveling in empty space with which the poem opened. Unraveling the sources used to formulate this contrast takes us deeper into the Cambridge setting for the poem's composition and demonstrates Empson's idiosyncratic approach to the intellectual resources around him.

"Letter I" was published in November 1928, by which time Lee had embarked on the second stage of his classical studies and Empson was beginning new studies in English literature. Part II of the Classical Tripos required students to specialize, and Lee had chosen ancient philosophy. This choice would have brought him under the tuition of Francis Cornford, one of the more progressive faculty members and a key participant in a group sometimes referred to as the "Cambridge Ritualists," who aimed to show that primitive rituals lay at the basis of ancient Greek society and thought. Well networked in Cambridge—his wife, the poet Frances Cornford, was Charles Darwin's granddaughter—Cornford cared as much about the unspoken assumptions shaping contemporary academic life as he did about the totemic rites underlying Greek philosophy. Cambridge University in the 1920s was tribal, too, in its way, and Cornford sought to puncture stuffy traditions with a more humane and flexible approach to learning. He was an inspiring teacher who viewed classical study as a source of constantly changing insights rather than a body of fixed knowledge to be foisted on successive generations. Addressing the newly founded Cambridge Classical Society in 1903, he compared their subject to cosmology in its need for constant revision: "As the philosophy of every new age puts a fresh and original construction on the universe," he observed, "so in the classics scholarship finds a perennial object for ever fresh and original interpretation."[78]

Cornford's books and lectures gave students in twentieth-century Cambridge a frisson of contact with the alien, visceral magic underlying rational thought and civilization. In the deep unity of tribal existence, there could be no division between body and mind, or society and nature. Empson found the classical scholar's primitive force an irresistible metaphor, using it as a focus for the eclectic literary, scientific, and philosophical influences that he was sucking up, like a hot-blooded version of Philip Quarles, from his intellectual environment. Empson's 1935 note to "Letter I" reveals how he set out to blend Cornford with modern astronomy in order to express extreme separation, in contrast to the poem's cheerful opening: "The network without fish is empty space which you could measure, lay an imaginary net of co-ordinates over, opposed in verse 3 to the condition when two stars are not connected by space at all; these are compared to two people without ideas or society in common, hence with no 'physics' between them in what F. M. Cornford said was the primitive sense of the word."[79]

The idea that "physics" had once formed the basis of social order stayed with Empson long after his departure from Cambridge. Cornford used the word *physis* to denote the living force that gave tribal life its profound unity. Empson summarized the concept in a critical essay published in 1935: "The primitive Greeks invented Nature by throwing out onto the universe the idea of a common life-blood; the living force that made natural events follow reasonable laws, and in particular made the crops grow, was identified with the blood which made the members of the tribe into a unity and which they shared with their totem."[80] Unlike the ancients, the speaker and addressee in "Letter I" do not share lifeblood and have no common basis for making sense of anything. As Empson explained in his note, "Lacking a common lifeblood shared from one totem (showman because tragic hero) they are connected by no idea whose name is derived from 'physics.'"[81] That baffling sequence of "sacrament," "showman," and "physician" is Empson's personal code for a figure blending the power of a tribal totem with that of Christ: a tragic hero who stands outside society and thereby makes cohesion possible.[82]

The latest cosmology, as Sinclair's protagonist discovered in "The Finding of the Absolute" and Russell stressed in his *ABC of Relativity*, cannot be relied on to provide any such cohesive force. Empson's convoluted third stanza brings out the irony of modern "physics." The intensely technical nature of relativity meant that any authors engaging deeply with its themes would immediately cut themselves off from most of the reading public, as media coverage during the early 1920s had shown. Moreover, the theory itself offered metaphors for intense disconnection. The banished lover's predicament in the "Letter" is "too non-Euclidean" because, like a particle traveling faster than light, his passion has worked to "crumple up round itself the whole of space-time."[83] He is now wrapped up in his own private universe, closed off from any contact with others. Here the poem ended in its 1928 magazine version, yearning for the comforting void that had been so feared by the recipient of the "Letter" at the beginning—for the darkness that "gives light its place," carrying light signals through space. By the end of the third stanza, the speaker cannot send or receive messages, and he is trapped in a system of meanings all his own. His beloved's face seems to be everywhere now, a constant reminder of thwarted intimacy.

During the next few years, Empson pursued the theme of intense curvature, engaging more deeply with Eddington's astrophysical theories. By the time his *Poems* were published in 1935, "Letter I" had gained a fourth stanza, describing a star completely cut off from surrounding space. This was a notion that Eddington was not prepared to entertain, as his comments in *Stars and Atoms* indicate.[84] At a meeting of the Royal Astronomical Society in January 1935, he ridiculed a young mathematician named Subramanyan Chandrasekhar who had been investigating the fate of massive dense stars. This story, with Professor Eddington in the role of an elder authority who held back research into what eventually became known as black holes, has been told several times from different angles.[85] Not being bound by the hierarchy of the astronomical establishment, Empson was free to explore the more peculiar features of Einstein's universe on his own terms. The added fourth verse of "Letter I" leaves us with a puzzle, however: What relationship does the poet's disconnected star have to the astronomical objects that were finally named in the 1960s? Did the mathematical poet really predict a black hole, simply by playing word games in non-Euclidean space?

The Hottest and the Coldest Matter in the Universe

Within the study of celestial mechanics, the positions and movements of heavenly bodies were charted with increasing precision, while all knowledge of what went on inside the stars, or what they might be made of, remained strictly out of bounds.[86] Only God needed to know what made them shine. The new discipline of astrophysics obtruded the gases, sparks, and stinks of laboratory work into serene contemplation of distant bodies through telescopes, and for several decades this new discipline relied heavily on the garden-shed enterprise of amateurs and the willingness of theorists to pursue speculative paths.[87] It all began during the 1860s with the development of spectroscopy, which revealed that rays of light told tales about where they had come from. Like the flame of a substance burned in the laboratory, starlight could be passed through a prism, resulting in a spectrum that gave information about its source and the gases it had traveled through. Messages from the forbidden interior of a star could now be received on earth, but they needed decoding. Without knowing quite how substances might behave when heated and compressed well beyond terrestrial conditions, it was hard to develop a coherent model for the sources of stellar energy and their changes through the lifetime of a star.

Eddington made a vital contribution to the foundations of this new area of investigation, but his methods were controversial, entailing what Matthew Stanley has described as a "dramatic departure from the tradition of celestial mechanics."[88] A more conventional method, advocated by James Jeans, consisted of starting with known facts about the stars and following a strict process of deduction to determine their likely behavior. But during the first two decades of the twentieth century, there were too many unknowns for this method to yield new insights. Instead of searching for strict proof, Eddington devised provisional hypotheses and models, testing them out on the observational data that were available and pushing them to see where they broke down. Stanley has detailed the ways in which Eddington "skillfully and rapidly moved beyond what he could prove and simply attempted to advance the theory," pointing out that the "uncertainty of his foundations was justified at the end of his work when he would demonstrate that his theory was insensitive to variations in the basic parameters." If Jeans was the Sherlock Holmes of British astronomy, Eddington was its maverick Lord Peter Wimsey, using theory as a means to further investigation rather than treating it as something to be deduced from established facts. This daring approach was not, however, the mark of an eccentric genius. Eddington's approach to astrophysical research was closely allied to the "seeking" after divine illumination that he had learned as a young Quaker embarking on a career in science.[89] Stanley's study of the Quaker astronomer reveals how these values enabled him to produce usable models of stellar energy before exact knowledge about the behavior of the subatomic particles driving stellar processes was available.

The findings of Eddington's research were gathered together in his groundbreaking technical book *The Internal Constitution of the Stars* (1926). Here he suggested that it was not necessary to wait until the mysteries of subatomic energy had been resolved because the results of any calculation based on such knowledge would surely have to agree with the stellar properties that could already be observed. One could devise models of stellar energy to fit those properties without worrying about the exact energy source. He offered an analogy: "The amount of water supplied to a town is the amount pumped at the waterworks; but it does not follow that a calculation based on the head of water and diameter of the mains is fallacious because it evades the problems of the pumping station."[90] It was the stability of water supply and stars that allowed for more than one approach to the same problem: "The two modes of calculating the water supplied to a town may not agree; but in that case there will be a flood at the pumping station. Similarly in a star a disagreement would involve the blowing up or collapse of the star." Precisely how a star avoided exploding was its own affair, for the time being. Eddington's insight consisted of isolating the problem of energy supply, which could not yet be resolved, from that of energy output, which could be observed night after night: "We may thus proceed with our method of determining the expenditure of radiation by the star without reference to the supply of subatomic energy. How the star manages to accommodate its supply to balance its expenditure, and so avoid collapse or expansion, is an independent problem." The result of this bold method was a law that linked a star's mass to its luminosity, a discovery remarkable for being applicable to both types of stars: gaseous giants and dense dwarfs.

Eddington had inherited a picture in which the stars were divided into two broad groups. Each star was supposed to begin life as relatively cool and extremely large ball of gas, rising in temperature as its radius contracted. When it became too dense for any further increase in temperature, the star would enter the dwarf sequence, cooling toward extinction. *The Internal Constitution of the Stars* opened with a survey chapter, in which Eddington took the liberty of speculating about a star of giant size with the density of a dwarf. He listed three peculiar features of such a star: light unable to escape the star's gravitational field; the spectrum shifted out of existence; and space closing up around the star.[91] The physicist Kip Thorne, in his popular book *Black Holes and Time Warps: Einstein's Outrageous Legacy* (1994), has described this passage as "whimsical," echoing the criticism of Eddington's expository style that was voiced as early as 1923 on the pages of *Science Progress*.[92] Thorne dismisses the third consideration, that "space would close up round the star," as "typical Eddington hyperbole."[93] But perhaps the eminent astronomer intended hyperbole here. We may imagine him joking—with colleagues at the Royal Astronomical Society's super-elite Dining Club, perhaps—about the absurd prospect of astronomers being left "nowhere" by rogue stars. A footnote in *The Internal Constitution of the Stars* indicates that the Cambridge mathematician Harold Jeffreys had drawn Eddington's attention to a relevant passage in the *Système du Monde* of Pierre-Simon Laplace, published in 1796.[94] Laplace had urged that a star with the density of the earth and a diameter two hundred and fifty times that of the sun "would not, in consequence of its attraction, allow any of its rays to arrive at us" and had concluded that "the largest luminous bodies in the universe may, through this cause, be invisible." Eddington offered the quote in a teasing manner: "Lest this argument should be regarded by our more conservative readers as ultra-modern," he remarked, "we hasten to add that it is to be found in the writings of Laplace." Einstein's theory was well established by 1926, but there were still prominent astronomers and physicists who held out hopes for an alternative to four-dimensional space-time. Well used to encountering such views at scientific meetings, at college dinners, and in correspondence, Eddington could not resist alarming this sector of his readership with a moment of Einsteinian excess, not suspecting for a moment that it might subsequently find a place in serious theorizing.

In the popular version of this discussion, published the following year in *Stars and Atoms*, Eddington left out the quotation from Laplace and concluded his commentary on hypothetical large dense stars by saying that it was only the tendency to leave the rest of the universe "nowhere" that made these objects absurd.[95] For that book's more general audience, he had to be clearer about which features of Einstein's theory were established fact and which were fanciful musings, and there was nothing to be gained by goading these readers into fourth-dimensional discomfort. The large dense star was allowed to remain, however, perhaps because it nicely encapsulates the modeling spirit of Eddington's researches. Throughout his technical work, extreme conditions are invoked as a way of discovering the limits on what is possible. Eddington worked by narrowing possibility down from impossibility, rather than broadening it out from the incontrovertible. But readers of popular science have a habit of deliberately running their thoughts against the grain of a text's priorities, and there will always be some who have the confidence or curiosity to skip between popular and more technical modes.[96] Perusing *The Internal Constitution of the Stars* alongside Eddington's popular lectures, Empson found this speculative approach to stellar mysteries a fitting corollary to his own problems. The astronomer's puzzling over how the stars were kept in equilibrium furnished the poet with further resources for thinking about the rejected lover's predicament. In contrast to astrophysical treatises, however, desperate love poems cannot escape so readily from the extreme conditions that they choose to evoke.

There was one particular type of star that evaded Eddington's theoretical grasp: the "mysterious" white dwarf.[97] Most stars, including our sun, were presumed to end their days in this state. White dwarf stars were baffling because they sent unreasonable messages to earth via the spectroscope. The companion of Sirius, for example, was advertising itself to astronomers as having a mass "about equal to the sun" and a radius "much less than Uranus."[98] These measurements entailed a density of "about a ton to the cubic inch," widely held to be preposterous. In *Stars and Atoms*, Eddington presented "The Story of the Companion of Sirius" as a detective mystery.[99] In reply to this star's claim that "a ton of my material would be a little nugget that you could put in a match box," astronomers had at first been inclined to retort, "Shut up. Don't talk nonsense."[100] Allowing that extreme density might be physically possible, Eddington articulated the heart of the mystery. Stars enter the white dwarf stage toward the end of their lives and must move on from it in order to die. In *The Internal Constitution of the Stars*, Eddington remarked that he could not see "how a star which has once got into this compressed condition is ever going to get out of it."[101] Close packing of matter is possible only when the outer electrons have been stripped away, and this would require the white dwarf to have a very high temperature indeed. As it cooled, the highly compressed star would have to lower its density, and to achieve this it would have to expand. A significant amount of energy would be needed to push its matter outward against gravity, but as Eddington observed, astronomers could "scarcely credit the star with sufficient foresight to retain more than 90 per cent. in reserve for the difficulty awaiting it." Bemused, he concluded that "the star will be in an awkward predicament when its supply of subatomic energy ultimately fails. Imagine a body continually losing heat but with insufficient energy to grow cold!"

An appendix to *Stars and Atoms* gave "a further instalment" to the companion of Sirius "detective story."[102] One suggested solution to the white dwarf's predicament had been that "subatomic energy will never cease to be liberated until it has removed the whole mass—or at least conducted the star out of the white dwarf condition."[103] But Eddington rejected this solution as being too much like "the device of a novelist who brings his characters into such a mess that the only solution is to kill them off." Recent developments in quantum theory—specifically, the "new statistics of Einstein and Bose and the wave-theory of Schrödinger"—had, he said, offered a more satisfying resolution of the problem. Eddington's colleague Ralph Fowler had suggested that wave properties of matter might provide a very dense star with its emergency energy supply. It was precisely this kind of collaboration between atomic theory and astronomy that helped to consolidate astrophysics as an exciting and rich discipline. As Eddington put it, the "white dwarf appears to be a happy hunting ground for the most revolutionary developments of theoretical physics." Fowler had taken his cue from hydrogen atoms that have their electron jammed up close against the central proton. According to quantum theory, such an atom cannot radiate, but its electron moves with high kinetic energy. This theory could be applied to a star in which the atoms were densely packed. As Eddington explained: "What is its temperature? If you measure temperature by radiating power its temperature is absolute zero, since the radiation is nil; if you measure temperature by the average speed of molecules its temperature is the highest attainable by matter. The final fate of the white dwarf is to become at the same time the hottest and the coldest matter in the universe."[104] The difficulty had been "doubly solved." Eddington's description of the fate devised under this new theory is a perfect contradiction: "Because the star is intensely hot it has enough energy to cool down if it wants to; because it is so intensely cold it has stopped radiating and no longer wants to grow any colder." This, he concluded, was "believed to be the final state of the white dwarf and perhaps therefore of every star," including our sun. A star in this condition would be invisible: like a hydrogen atom with its electron jammed close to the proton, it would "give no light." Here was the perfect metaphor for Empson's inexpressible passion.

Stellar Buffoonery

Victorian debates over "heat death" (the winding down of the universe into thermodynamic equilibrium) and the age of the earth and solar system had given the fate of our sun a prominent place in popular and literary imagination.[105] Devising a fourth stanza for his "Letter" poem, Empson used the sun's updated astrophysical destiny to intensify the pain of unrequited love:

Our jovial sun, if he avoids exploding
(These times are critical), will cease to grin,
Will lose your circumambient foreboding;

Loose the whole radiance his mass can win
While packed with mass holds all that radiance in;
Flame far too hot not to seem utter cold
And hide a tumult never to be told.[106]

J. B. S. Haldane's popular writing, particularly the concluding piece from *Possible Worlds*, jostles with Eddington here. In "The Last Judgment," Haldane gleefully describes James Jeans's suggestion that the sun might divide, roasting mankind and boiling the sea.[107] The previous generation had feared it would cool down within a few million years, while modern physics was more cheerfully predicting a million million years at least. Still, there was a chance that the sun would burst before it ran out of energy, "expanding enormously, giving out a vast amount of heat, and then dying down again."[108]

Empson's "jovial" sun (named after the chief Roman god Jove, or Jupiter, father of Mars) may "avoid exploding," a phrase that blends Eddington's daring to let the star manage its own energy supply with Haldane's deliberately casual tone. In his review of *Possible Worlds*, Empson relished this teasing quality in Haldane's prose: "the scientist, innocent, cocksure, pretending not to notice the elaborate, long-ripened toes he treads on, answers any eternal questions that turn up, and hobnobs briskly with the stars."[109] But the poet also recognized that this kind of writing could wear thin if it persisted in disregarding human interest. Haldane's essays were saved, he concluded, partly by their wit and noble sentiments, but also by the "hints continually thrown out in just the right place that the author, being a man of culture, really knows better and is saying this or that merely to chivvy somebody."[110] Incorporated into a love poem, Haldane's chivvying style helped Empson to universalize his feelings, establishing a tension between the attempt to be brisk and sensible and the pull of unbearable passions. As the poem reaches its conclusion, the compelling metaphor of Eddington's "ultra-modern" dark star makes it harder to maintain Haldane's insouciance. Empson's note explained how he was using the science: "A big enough and concentrated enough star would, I understand, separate itself out from our space altogether. Verse 4 describes a similar failure of communication which may in the end happen to the sun; *your circumambient foreboding* is 'the empty space round him which connects us to him and which you fear.'"[111] The space between earth and sun, previously filled with hopeful radio messages, may be lost if the star somehow manages to become very large and dense, ceasing to "grin" or shine as its rays fail to escape from the intense gravitational field. But this was hardly the fate that astronomers had in store for our sun. What Empson has done here, in one of his characteristic imaginative raids, is to bring Eddington's hypothetical massive dense star together with his account of what may happen to our sun as it becomes a white dwarf at the end of its life. Wishing to maximize the poetic opportunities afforded by Eddington's detective story, Empson has also incorporated the rejected novelistic solution, in which the liberation of subatomic energy will remove the star's whole mass or "loose the full radiance his mass can win." This scientifically inappropriate melding of two kinds of "invisibility" leads to the creation of something resembling a black hole in a poem that was complete by 1935. The fictive dense giant, "packed with mass" and holding "all that radiance in," is conjoined with the white dwarf's temperature paradox to produce a powerful couplet giving Einsteinian expression to a love that "dare not speak its name": "Flame far too hot not to seem utter cold / And hide a tumult never to be told."

The choice of "tumult" for the final line contributes to a satisfying rhythm, but it also pays tribute to the humanizing energy of Eddington's science writing. Even in his technical books, the professor could not resist giving stellar lives a dramatic, violent edge. Between a section headed "Radiation Pressure" and another headed "Opacity of Stellar Material" in *The Internal Constitution of the Stars*, he allowed the excitement of this new conjunction between physics and astronomy to run riot in his prose:

> The inside of a star is a hurly-burly of atoms, electrons and aether waves. We have to call to aid the most recent discoveries of atomic physics to follow the intricacies of the dance. We started to explore the inside of a star; we soon find ourselves exploring the inside of an atom. Try to picture the tumult! Dishevelled atoms tear along at 50 miles a second with only a few tatters left of their elaborate cloaks of electrons torn from them in the scrimmage. The lost electrons are speeding a hundred times faster to find new resting-places. Look out! there is nearly a collision as an electron approaches an atomic nucleus; but putting on speed it sweeps round it in a sharp curve. . . . Then comes a worse slip than usual; the electron is fairly caught and attached to the atom, and its career of freedom is at an end. But only for an instant. Barely has the atom arranged the new scalp on its girdle when a quantum of aether waves runs into it. With a great explosion the electron is off again for further adventures. Elsewhere two of the atoms are meeting full tilt and rebounding, with further disaster to their scanty remains of vesture.
>
> As we watch the scene we ask ourselves, Can this be the stately drama of stellar evolution? It is more like the jolly-crockery-smashing turn of a music-hall.[112]

Relishing the music hall scene as an alternative to "stately drama" of the spheres, Eddington signals his ability to set conservative scientific traditions to one side. A scientist of the Ronald Ross stripe would hardly be invoking scalps, scanty vesture, and jolly crockery smashing in his perusal of the heavens. Eddington was not afraid to abandon strictly deductive reasoning in his pursuit of pragmatic models for stellar behavior, nor was he afraid to blur the lines between "popular" and "specialist" writing. The Quaker astronomer's combination of irreverence and passion provided the ideal source material for poems that aimed to give personal dilemmas a more universal form without diluting their intensity.

Empson had been trained as a Cambridge mathematician, and though his upper second class result was disappointing (no wrangler status), he could not have entirely escaped the various courses relevant to stellar energy and relativity.[113] Was there a mathematical or physical insight at work in the connection that he forged between the white dwarf's inability to shine and the dark star's entrapment? Chandrasekhar's discovery of a "limiting mass," a size above which a dying star might undergo complete gravitational collapse, was announced at a meeting of the Royal Astronomical Society in January 1935.[114] Eddington ridiculed these results before the assembled astronomers, and in the following year he famously dismissed the prospect as "stellar buffoonery."[115] If Empson had been following discussions in astronomy closely, he might have picked up on the impending development of stellar fates before completing his fourth verse. In a paper about white dwarfs published in 1931, for instance, Chandrasekhar had hypothesized about highly collapsed stars that could not be seen.[116] But there is no evidence to suggest that the poet was pursuing this as a mathematical or astrophysical problem, nor was he sufficiently trained in those disciplines to do so. Empson's dying sun, separated out from the rest of space, is not a "black hole" in the astrophysical sense. What, then, are we to make of his poem's premature description of a star that collapses and disappears into its own private space-time? Is it merely coincidence of poetic and scientific imagination? The conclusion to "Letter I" picks up on two conditions that were marked out by Eddington as "impossible" in some way: a massive dense star with space closed up around it, and a white dwarf with "insufficient energy to grow cold." It was the outlawed status of these hypothetical objects that made them suitable to help transmute the poet's passion for a heterosexual male friend into something beyond the personal. The result has a predictive power because Empson was free to explore ideas that remained out of bounds to members of the Royal Astronomical Society under Eddington's authority.

The fact that other theorists eventually succeeded in pushing astrophysical research toward the forbidden "buffoonery" must have delighted Empson, who certainly kept an eye on subsequent developments in this area. Among his papers is "The Search for Black Holes" by Kip Thorne, torn from a 1974 issue of *Scientific American*.[117] Thorne begins with a frank acknowledgment of this topic's preposterous nature: "Of all the conceptions of the human mind from unicorns to gargoyles to the hydrogen bomb perhaps the most fantastic is the black hole."[118] This peculiar object, he concedes, "seems much more at home in science fiction or in ancient myth than in the real universe." Drawn to its fantastic features about thirty years before Einstein's "outrageous legacy" had been settled, Empson created a home for the forbidden dark star in his "Letter" poem. Here it could be accommodated alongside mythical and science fiction themes that he had absorbed from Cambridge studies of primitive Greek religion and from popular excitement about life on other worlds.

As Empson's description of walking into a scene from *The Sleeper Awakes* during the General Strike suggests, there is more to prediction than physical or theoretical detail. It was the genius of Wells, the poet suggested, to have "forecast not merely the invention of the machine and its subsequent political importance but also the way Englishmen would feel about it in its political use."[119] What Empson has done in "Letter I" is comparable to what he saw in Wells's depiction of workers jostling around the source of broadcast propaganda, and in Donne's metaphysical conceits giving lovers sovereignty over new worlds. All three authors explored the social and political implications of a new development in science or technology.[120] The radically separated dark star at the conclusion of his poem is not a black hole in the astrophysical sense that was eventually established in the 1960s. But the poet's "too non-Euclidean predicament" in "Letter I" does respond acutely to the context of science popularization in 1920s Britain, recognizing cosmic speculation as an activity through which elite authority and mass interests may be renegotiated. Empson's poems on Einstein themes associate the curvature of space-time with concerns about the tendency for specialists to become cut off from the mass of readers. They also challenge the celebration of individual imagination that arose in response to popular cosmology, exemplified in *Armchair Science* coverage. If every reader has the right to speculate about the cosmos on his or her own terms, then readers also risk becoming lost from their fellows—slipping through a crack in culture to end up "nowhere," inhabiting a specialism of their own concoction. Like Aldous Huxley's early satirical writings, Empson's astronomy love poems place their energies in the service of elaborating the lack of secure ground for modern readers, rather than devising solutions.

Conclusion: Dreaming the Future

By bringing together news coverage, expository writing, popular fiction, and esoteric poetry on Einstein themes, we have gained a richer sense of what "relativity" came to mean in Britain during the early twentieth century. The new theory of space, time, and gravitation provoked a great deal of frustration and anxiety, but it also proved useful and entertaining to mathematical innocents, who exploited the new cosmology to diverse ends. The reception of relativity takes us to the heart of the contradictions that characterized early-twentieth-century British culture. By listening in on talk of bent light rays and time as a fourth dimension, we are better able to sense the continued dominance of certain Victorian preoccupations—such as progress and degeneration, absolute and relative values, and tension between elite learning and popular entertainment—while perceiving how these preoccupations were given a new intensity—and fragility—as people came to terms with the impact of war, shifting sex and class relations, new technologies of entertainment, faster communication, and expanding consumer culture. We may compare Einstein himself, who extended the crowning achievement of nineteenth-century physics—James Clerk Maxwell's electromagnetic theory—to the point where classical laws broke down, with audiences for relativity who used the new space and time to explore cracks in nineteenth-century culture, registering the decline of Liberal government, questioning access to knowledge, renegotiating sex relations, coming to terms with war loss, and colluding with media power. While physicists pondered the loss of an absolute "now," British consumers tried to work out which primitive traits, Victorian achievements, and contemporary phenomena should be celebrated or rejected.

The "revolution in science" proved a rich and versatile cultural resource in Britain between the wars, its significance extending far beyond the Victorian association of "relativity" with "relativism." A full picture of how this worked emerges only when the different genres of writing are brought together. The potential for questioning and reimagining the world, offered to readers through newspaper coverage, popular fiction, popular science, and poetry, is a collaboration across the cultural spectrum. The practical need to think about distinct audiences and their interests can sometimes obscure the power and complexity of science in culture as an extended process of negotiation. This process does not have sharp boundaries, and as we have seen, it may not reach a resolution before the issues under negotiation are picked up in an adjacent area of communication. The literary critic and historian of science Ralph O'Connor has observed that the study of literature and science remains somewhat divided: "Most literary scholars still use scientific writings chiefly as intellectual 'contexts' for illuminating . . . novels, plays, poetry, and essays," he notes, "while most historians of science treat novels, plays, and poetry as second-order cultural background for the explication of scientific debates."[1] O'Connor also suggests that greater attention to the complexities of genre and closer engagement with literary studies will enhance the study of popular science.[2] The example of relativity in Britain suggests one way to overcome this divide between literary and historical approaches, displaying a trajectory within which each genre of writing is seen to play a specific role. Close analysis of textual detail adds up to a story that is keyed into British politics and scientific debates of the day.

Securely interlocked readerships are not necessary to the argument here, which is about different forms of expertise being applied to a shared problem. It's unlikely that Eddington picked up the *Red Magazine* while preparing his early expositions of relativity, or that former subscribers to *Conquest* had 12s. 6d. to spend on a first edition of *The Nature of the Physical World* in 1928. What can be traced and compared across these texts is the treatment of physical abstraction in relation to human interest and the renegotiation of what is held in common. These themes suggest a starting point for further work on the reception of Einstein in different national contexts.[3] Will the narrative arc from class conflict fueled by newspaper science to participation enabled by idealist astronomers be the same in Latin America, in Eastern Europe, in India? Has the story been inflected through mass-market fiction and neo-metaphysical poetry in other countries, or have other cultural forms been used to negotiate meanings for relativity? There is plenty still to find out about what happened around the world when light was caught bending in 1919. There is also more work to be done on the British context. The engagement of canonical literary authors such as Virginia Woolf and Aldous Huxley with Eddington, Jeans, Russell, Sullivan, and Whitehead has already received considerable attention from scholars, but the inclusion of more ephemeral material from newspapers and popular magazines may reveal greater depth and complexity in the literary uses of science, as it has in the cases of Sayers and Empson. And there are plenty more pulp authors to be explored whose writings represent a virtually untapped archive of popular opinion, albeit one inflected through commercial pressures.

"It is purely abstract science."[4] So Einstein is reported to have said in protecting his work from entanglement with religion or politics, and many scientists today would agree with regard to their own research. Relativity is a good example of how new discoveries in science will inevitably be inflected through the politics of the day, and it teaches us that such encounters are far from simplistic. The new cosmology proved very useful in helping to suppress any Bolshevist sympathies that might have been breeding in the hearts of British readers. At the same time, labor interests were used to challenge the scientific establishment. The result was a curious tension at the intersection of science and politics. Newspaper jokes linking relativity with social revolution suggested that Bolshevism was as far-fetched or irrelevant to daily life as a warped ray of light or a voyage in the fourth dimension. The Bolshevism link was also used to suggest that scientists had reached absurd new heights of abstraction, removing themselves from the realm of common sense: a complaint that potentially associated social revolution with the demand for more practical, more accessible science. By casting scientists as absurd revolutionaries, a largely conservative popular press used socialism and relativity to neutralize each other, whereas these two features of the postwar cultural landscape might easily have entered an explosive relationship of mutual reinforcement. The prevalence of jokes on this topic implied that such reinforcement was a real possibility, even if writers in support of socialism (such as Bertrand Russell) often chose to maintain a separation between science and politics. Negation of labor interests continued on the pages of popular science and fiction magazines, albeit in a more diffuse manner. The demand for popular access to Einstein's universe was denied by science editors who wished to foster appreciation of British science while maintaining an authoritative distance between increasingly specialized scientific expertise and "the man in the street." Various models of work were deployed in an attempt to make science relevant to postwar consumers: King Sol versus trade unions in *Conquest*; elite professionals with a supportive audience in *Discovery*; entertainment and individualism in *Armchair Science*. The less participatory the model, the more frustration and disappointment lay in store for readers. Comprehension of relativity took on a symbolic quality comparable to the living wage in socialist arguments, representing a wider struggle for shared knowledge. But when more participatory contexts for popular relativity did emerge in the late 1920s and early 1930s—for example, with Eddington's exposition and the "melting pot" philosophy of *Armchair Science*—those contexts were far from radical or collective. Access to Einstein's universe was granted, but on grounds of entertainment and creativity of mind, rather than "work."

For all the diverse efforts of their contributors, popular science magazines lacked both the expository techniques and the supporting ideological context conducive to fruitful negotiation between the common-sense labor of readers and the elite owners of mathematical abstraction. Authors of pulp fiction, writing for magazines owned by conservative interests, were deft in their handling of this impasse. The conventions of supernatural storytelling offered an ideal form through which to remedy the failure of newspaper and magazine exposition without espousing socialism. By adopting elements of the uncanny tale rather than pure science fiction as the preferred genre for Einstein-themed stories, authors such as Coutts Brisbane and A. E. Ashford helped break the underlying association between social and scientific revolutions, opening up relativity to more diverse uses. By the late 1920s Eddington was able to argue at length against a parochial outlook on the universe without any risk of alignment with socialism. His exposition radiated so strongly with individualism that interpreters such as Sayers and Empson were compelled to question the intake of popular physics books that threatened to cut readers off from one another and from the rest of society. At first sight, the melodramatic responses to relativity articulated by Sayers and Empson appear to chime with the concern of J. J. Thomson about every individualist ending up in "a universe of his own." But there is something oddly necessary about the four-dimensional discomforts elaborated through Sayers's cosmic bachelors and Empson's astronomy love poems. Science is being used to dramatize a sense of isolation and disconnection, lending focus to postwar social concerns about intimacy and identity. By enrolling the new cosmology in the exploration of these concerns, the authors of these stories and poems also perform a service for science communicators, raising and intensifying questions about how popularization intersects with other challenges in readers' lives.

The case of relativity in Britain suggests that expository writing can either be restricted to "pure science," reaching a restricted audience, or mix science with pressing issues of the day and be relevant to wider audiences. This might be a depressing conclusion for some, if not so surprising to others. But it also represents an opportunity to rethink the functions of science communication in relation to other cultural activities. The special contribution of science communicators may be understood as one of posing or intensifying problems—problems that are received not just from science but also from public debate in the media, for instance. Individual communicators and institutions may choose how far they wish to travel down this path, though even the most restricted type of exposition cannot shield itself from the "other knowledge, other questions" invoked by its audience. The failure of popularized science to resolve the social tensions that it sharpens makes it potentially more radical as a genre than newspaper coverage or some forms of popular entertainment (today's equivalent of the *Red Magazine* might be found among TV serials, graphic novels, and films engaging contemporary science themes). Its power lies in the questions provoked, rather than in any answers that can be given. Other types of writing or storytelling are needed to help exploit this potential, for the expositors themselves may not be best placed to exploit the disjunction between wider cultural needs and scientific insight on which their encounter with a wider public thrives. In the 1920s and 1930s, the richest forms of exploitation came from authors who were able to mix elite learning with mass culture: Wimsey's encounter with Duckworthy over the question of a fourth dimension in Sayers's detective fiction, for example, or Empson's mobilization of the Martian wireless theme alongside Cornford's theory of ancient religion in "Letter I." Pulp stories, by contrast, were more concerned to reaffirm boundaries of class and leisure interest.

The capacity for mixing across cultural registers helps to explain the remarkable success of a book that propounded a heavily abstruse theory of time, devised by the pioneering aeronautical engineer J. W. Dunne. *An Experiment with Time* was published in 1927, the year before Eddington's Gifford Lectures appeared in print. That book and its sequel, *The Serial Universe* (1934), received attention in philosophical and scientific journals as well as appealing to wider public audiences for several decades. Dunne put forward an elaborate philosophical scheme for establishing that the time of our waking experience unfolds within a second order of time, and so on with infinite regression. His arguments built toward the existence of a group mind, uniting all individuals in a "greater now" that granted immortality of the soul, a theme developed in his subsequent writings. Both books were reviewed, critically but with careful attention, in *Discovery* and *Nature*. The credibility of Dunne's philosophy appeared to be further confirmed by the inclusion, from the second edition of *An Experiment with Time* onward, of an extract from a letter by Arthur Eddington dated February 1928, beginning with the words "I agree with you about 'serialism.'" Readers' experience of the Einstein sensation would have made Dunne's narrative attractive: a man of practical affairs had devised his own theory of time as an alternative to Einstein's relativity.

An Experiment with Time, revised in 1929 and 1934, invites readers to participate in a large-scale experiment to help prove serialism by recording evidence that they have strayed into the future while dreaming. Instructions for how to record and analyze dream data are provided, alongside socially situated accounts of reading such as daydreaming about a fishing expedition in a friend's trout stream while perusing a small-print article in the *Encyclopaedia Britannica*, receiving the *Daily Telegraph* while abroad, or settling down for a nap in a London club with a popular novel in hand. These vignettes about reading not only serve specific expository functions, but also establish the author's milieu while creating sympathetic links to readers of a different class who might themselves have struggled with an encyclopedia entry or enjoyed reading adventure stories. The wide appeal of Dunne's serial time may be gauged from the number and range of authors who took inspiration from *An Experiment with Time*, whose influence extended far beyond the "Golders Green to Teddington" swathe of London's garden suburbs touched by "Einstein, Jeans, and Eddington." While Dunne's influence on various science fiction, fantasy, and canonical literary authors has been acknowledged, no comprehensive survey has been published, nor has the texture of his exposition been subject to sustained analysis.[5]

The Dunne phenomenon will repay exploration from various angles, but it also serves as a fitting conclusion to our discussion of the reception of relativity in Britain, for it recapitulates each stage in the trajectory from alienation to participation. *An Experiment with Time* delivered precisely what readers of *Conquest* had craved: personal access to the time dimension. Whereas popular fiction had warned readers away from getting too close to the new space and time, Dunne offered redemptive powers to all of humanity, to be accessed through training in awareness of a "greater now." The excessive individualism that came to be associated with bourgeois enjoyment of popular physics already had its embryonic answer, in 1927, in Dunne's appeal to interconnected souls at the conclusion of his book. Dunne gave the public what Einstein couldn't: a way into the ethical consequences of reimagining space and time. And the many storytellers who engaged with serial time subjected Dunne's utopian vision of human potential to sustained questioning.

As Einstein's relativity began to take up its postponed role in science fiction tales of light-speed travel and temporal paradox, *An Experiment with Time* continued to offer an uncanny resource through which to explore potentially hazardous experience of disrupted time. The Scots barrister and prolific mystery author John Buchan seized on this potential at the heart of serial time in his novella *The Gap in the Curtain* (1932). Five men are trained to predict the future by entering a specially induced state of consciousness and gazing at a page of the *Times*, which allows them to discern what will be reported one year later. Lives are wrecked by foreknowledge of various kinds until the final character manages to elude his own announced death, bolstered by his fiancée's healthy skepticism toward the twinned cultural authority of newspapers and scientists. Buchan recognized that Dunne's exposition offered a way to dramatize a problem that had lurked beneath coverage of relativity from the start: what happens to the individual's capacity for free will, speculation, and imagination when science and the media are locked into mutual self-reinforcement? This, in the end, is the most powerful form of mutual enrollment that takes place through science communication, and it is one that cannot easily accommodate other interests. Dunne's *Experiment with Time* inspired many more literary projects than Einstein or Eddington did, and they are spread right across the cultural spectrum, from science fiction and fantasy and middlebrow fiction to modernist esoterica. Among these projects will be examples like those of Sayers and Empson, in which alternative forms of enrollment emerge.

Throughout this book I have used the term "everyday life" in a simple way, but it is worth being alert to complexities in its use. It is all too easy to end up colonizing the "everyday" for the purposes of scholarship, thereby reinforcing the gulf that is supposed to be bridged—just as relativity expositors did when they used swimmers, trains, and clocks to underline the abstraction of relativity from familiar experience. In a 2011 symposium review of Richard Staley's *Einstein's Generation* (2008), Lewis Pyenson identifies a next step for research in the history of physics: "Staley has contributed significantly to a rich body of literature about physicists in the laboratory and at the writing desk. Attention may now profitably turn to the daily routine of physicists, including meals, sleeping arrangements, and laundry; recreations, including music and athletics; wealth and health; political and religious exercise; the newspapers they took and the novels they read."[6] There is a choice to be made about how we pay this kind of attention. Pyenson suggests that such an approach "is required to establish the place of the new physics of relativity in the High Modern Age, for our inclination, following 40 years of scholarship in social history of ideas, is to imagine that physicists adjusted their thought to make abstract formulations fit with the spirit of the times, as they perceived it." This practice would effectively align scientific theorists with producers of canonical high modernist works, such as T. S. Eliot. I have shown that the richness of relativity as a cultural resource emerges only when we locate it amid the melee of highbrow, lowbrow, and middlebrow modernity that mathematically innocent readers were faced with as they worked to renegotiate the conditions of everyday life between the wars.

Notes

EPIGRAPH

Quoted by John Haffenden in his introduction to William Empson, *Essays on Renaissance Literature*, vol. 1, *Donne and the New Philosophy*, ed. John Haffenden (Cambridge: Cambridge University Press, 1993), 4.

INTRODUCTION

1. On the reception and appropriation of evolution theory, see, for example, Gillian Beer, *Darwin's Plots: Evolutionary Narrative in Darwin, George Eliot and Nineteenth-Century Fiction* (London: Routledge & Kegan Paul, 1983); Peter Bowler, *Charles Darwin: The Man and His Influence* (Cambridge, MA: Blackwell, 1990); Thomas Glick, ed., *The Comparative Reception of Darwinism* (Austin: University of Texas Press, 1974); George Levine, *Darwin and the Novelists: Patterns of Science in Victorian Fiction* (Cambridge, MA: Harvard University Press, 1988); James A. Secord, *Victorian Sensation: The Extraordinary Publication, Reception, and Secret Authorship of* Vestiges of the Natural History of Creation (Chicago: University of Chicago Press, 2000); Eve-Marie Engels and Thomas Glick, eds., *The Reception of Charles Darwin in Europe* (London: Continuum, 2008).
2. On media self-reflectivity in the 1830s, see Secord, *Victorian Sensation*, 30.
3. "Einstein on His Theory: Time, Space and Gravitation, The Newtonian System," *Times*, November 28, 1919, 13–14.
4. *The Mathematical Theory of Relativity*, by A. Kopff, review by S. B., *Science Progress* 18, no. 72 (1924): 647.

5. "Bent Light Puzzle—Euclid Not 'Caught Out,'" *Daily Mail*, November 8, 1919, 5; "Lights All Askew in the Heavens," *New York Times*, November 10, 1919, 17.
6. L. G. Brazier, "What Is the Use of Einstein?," *Conquest* 3, no. 3 (January 1922): 123–24. Emphasis in original.
7. Arthur R. Burrows, "Juggling with Air," *Conquest* 1, no. 3 (January 1920): 107; Charles Davidson, "Weighing Light," *Conquest* 1, no. 3 (January 1920): 125.
8. Arnold Bennett, *The Evening Standard Years: "Books and Persons" 1926–1931*, ed. Andrew Mylett (London: Chatto and Windus, 1974), 36.
9. Secord, *Victorian Sensation*, 532.
10. My emphasis on mass culture complements previous studies of literature and physical science that have focused on canonical and/or high modernist authors: Gillian Beer, "Eddington and the Idiom of Modernism," in *Science, Reason, and Rhetoric*, ed. Henry Krips, J. E. McGuire, and Trevor Melia (Pittsburgh: University of Pittsburgh Press, 1995), 295–315; Beer, "'Wireless': Popular Physics, Radio and Modernism," in *Cultural Babbage: Technology, Time, and Invention*, ed. Francis Spufford and Jenny Uglow (London: Faber, 1996), 149–66; Beer, *Wave, Atom, Dinosaur: Woolf's Science* (London: Virginia Woolf Society, 2000); David Bradshaw, "The Best of Companions: J. W. N. Sullivan, Aldous Huxley, and the New Physics [Parts 1 and 2]," *Review of English Studies*, n.s. 47, no. 186 (1996): 188–206; no. 187 (1996): 352–68; Michael Whitworth, "'Pièces d'identité': T. S. Eliot, J. W. N. Sullivan and Poetic Impersonality," *English Literature in Transition* 39, no. 2 (1996): 149–70; Whitworth, *Einstein's Wake: Relativity, Metaphor, and Modernist Literature* (Oxford: Oxford University Press, 2001); Daniel Albright, *Quantum Poetics: Yeats, Pound, Eliot and the Science of Modernism* (Cambridge: Cambridge University Press, 1997); Daniel Brown, *Hopkins' Idealism: Philosophy, Physics, Poetry* (Oxford: Clarendon Press, 1997); Sarah Cain, "The Metaphorical Field: Post-Newtonian Physics and Modernist Literature," *Cambridge Quarterly* 28 (1999): 46–64; Ann Banfield, *The Phantom Table: Woolf, Fry, Russell, and the Epistemology of Modernism* (Cambridge: Cambridge University Press, 2000); Daniel Tiffany, *Toy Medium: Materialism and the Modern Lyric* (Berkeley: University of California Press, 2000); Holly Henry, *Virginia Woolf and the Discourse of Science: The Aesthetics of Astronomy* (Cambridge: Cambridge University Press, 2003); Anna Henchman, "'The Globe we groan in': Astronomical Distance and Stellar Decay in *In Memoriam*," *Victorian Poetry* 41 (2003): 29–45; Henchman, "Hardy's Stargazers and the Astronomy of Other Minds," *Victorian Studies* 51 (2008): 37–64; Pamela Gossin, *Thomas Hardy's Novel Universe: Astronomy, Cosmology and Gender in the Post-Darwinian World* (Aldershot: Ashgate, 2007); Barri Gold, *Thermopoetics: Energy in Victorian Literature and Science* (Cambridge, MA: MIT Press, 2010); Candice Kent, "'How does the mind move to Einstein's physics?': Science in the Writings of Virginia Woolf and Mary Butts," in *Re-*

storing the Mystery of the Rainbow: Literature's Refraction of Science, eds. Valeria Tinkler-Villani and C. C. Barfoot (Amsterdam: Rodopi, 2011), 567–83. Bruce Clarke's *Energy Forms: Allegory and Science in the Era of Classical Thermodynamics* (Ann Arbor: University of Michigan Press, 2001) does refer to popular fiction alongside canonical works, while Laura Otis, "The Other End of the Wire: Uncertainties of Organic and Telegraphic Communication," *Configurations* 9 (2001): 181–206, explores telegraphy in British popular fiction, and Michael Whitworth, "The Clothbound Universe: Popular Physics Books, 1919–1939," *Publishing History* 40 (1996): 53–82, analyzes best-selling popular physics books. Studies of Einstein themes in American modernist literature have paid closer attention to the mass market; see note 16.

11. Gillian Beer, *Open Fields: Science in Cultural Encounter* (Oxford: Oxford University Press, 1996), 187.
12. On Victorian debates over relative and absolute values, see Christopher Herbert, *Victorian Relativity: Radical Thought and Scientific Discovery* (Chicago: University of Chicago Press, 2001).
13. Robert J. Strutt (Lord Rayleigh), *The Life of Sir J. J. Thomson* (Cambridge: Cambridge University Press, 1942), 203–4; G. K. A. Bell, *Randall Davidson, Archbishop of Canterbury* (Oxford: Oxford University Press, 1935), 1052; J. J. Thomson, *Recollections and Reflections* (London: Bell, 1936), 431; Arthur Eddington, *The Philosophy of Physical Science* (Cambridge: Cambridge University Press, 1939), 7; Philipp Frank, *Einstein: His Life and Times*, trans. George Rosen (New York: Knopf, 1947), 189–90; Ronald Clark, *Einstein: The Life and Times* (London: Hodder and Stoughton, 1979), 266–68; Max Jammer, *Einstein and Religion: Physics and Theology* (Princeton: Princeton University Press, 1999), 155; Peter E. Hodgson, "Relativity and Religion: The Abuse of Einstein's Theory," *Zygon* 38, no. 2 (2003): 393–409. See Peter Bowler, *Reconciling Science and Religion: The Debate in Early-Twentieth-Century Britain* (Chicago: University of Chicago Press, 2001), 102, for a brief discussion.
14. On relativity and modern art, see Linda Dalrymple Henderson, *The Fourth Dimension and Non-Euclidean Geometry in Modern Art*, 2nd ed. (Cambridge, MA: MIT Press, forthcoming); Henderson, "Einstein and 20th-Century Art: A Romance of Many Dimensions," in *Einstein for the 21st Century: His Legacy in Science, Art, and Modern Culture*, ed. Peter L. Galison, Gerald Holton, and Silvan S. Schweber (Princeton: Princeton University Press, 2008), 101–29; Gavin Parkinson, *Surrealism, Art, and Modern Science: Relativity, Quantum Mechanics, Epistemology* (New Haven: Yale University Press, 2008). On relativity and modernist literature, see Whitworth, *Einstein's Wake*; Clarke, *Energy Forms*; Randall Stevenson, *Modernist Fiction: An Introduction*, rev. ed. (Harlow: Prentice Hall, 1998; first published 1992); Betty Jean Craige, *Literary Relativity: An Essay on Twentieth-Century Narrative* (Lewisburg: Bucknell University Press, 1982).
15. "Why Nothing Stands Still," *Tit-Bits*, March 5, 1921, 7.

16. For example, from the *New York Times*: "Light and Logic," November 16, 1919, sec. 3, 1; "Jazz in Scientific World," November 16, 1919, sec. 3, 8. On the reception of Einstein in America, see Alan J. Friedman and Carol C. Donley, *Einstein as Myth and Muse* (Cambridge: Cambridge University Press, 1985); Lisa Steinman, *Made in America: Science, Technology and American Modernist Poets* (New Haven: Yale University Press, 1987); Eric White, "Advertising Localist Modernism: William Carlos Williams, 'Aladdin Einstein' and the Transatlantic Avant-Garde in *Contact*," *European Journal of American Culture* 28 (2009): 141–65.
17. Bernard Lightman, *Victorian Popularizers of Science: Designing Nature for New Audiences* (Chicago: University of Chicago Press, 2007), 218, 169.
18. On the pre-Einsteinian fourth dimension, its subsequent conflation with Einstein's relativity, and its uses in art and literature, see Henderson, *Fourth Dimension*; Henderson, "Einstein and 20th-Century Art"; Henderson, "Four-Dimensional Space or Space-Time? The Emergence of the Cubism-Relativity Myth in New York in the 1940s," in *The Visual Mind II*, ed. Michele Emmer (Cambridge, MA: MIT Press, 2005), 349–97; Elizabeth Throesch, "Charles Howard Hinton's Fourth Dimension and the Phenomenology of the Scientific Romances (1884–1886)," *Foundation: The International Review of Science Fiction* 99 (2007): 29–49; Throesch, "The Alice Books and the Fourth Dimension: Victorian Fantastic Spaces," in *Alice Beyond Wonderland: Essays for the Twenty-First Century*, ed. Cristopher Hollingsworth (Iowa City: University of Iowa Press, 2009), 37–52.
19. W. H. Auden and Louis MacNeice, *Letters from Iceland* (London: Faber, 1937), 103.
20. Thomson, *Recollections and Reflections*, 432–33.
21. On Eddington and Jeans, and on responses to their exposition, see Whitworth, "Clothbound Universe" and *Einstein's Wake*; Beer, "Eddington and the Idiom of Modernism," "'Wireless,'" and *Wave, Atom, Dinosaur*; Bowler, *Reconciling Science and Religion*; Bowler, *Science For All: The Popularization of Science in Early Twentieth-Century Britain* (Chicago: University of Chicago Press, 2009); Henry, *Virginia Woolf*; Elizabeth Leane, *Reading Popular Physics: Disciplinary Skirmishes and Textual Strategies* (Aldershot: Ashgate, 2007).
22. "The Run Home: An Einsteinist at Large" and "'A Fashion of To-Day,'" *Times*, December 2, 1919, 17.
23. "Light on the Bend," *Morning Post*, November 8, 1919, 7.
24. "The Fourth Dimension," *Morning Post*, November 11, 1919, 8.
25. On war neurosis, see Trudi Tate, *Modernism, History and the First World War* (Manchester: Manchester University Press, 1998).
26. On Cambridge responses to special relativity, see Andrew Warwick, "Cambridge Mathematics and Cavendish Physics: Cunningham, Campbell, and Einstein's Relativity, 1905–1911, Part 1: The Uses of Theory," *Studies in History and Philosophy of Science* 23 (1992): 625–56, and "Part 2: Comparing

Traditions in Cambridge Physics," *Studies in History and Philosophy of Science* 24 (1993): 1–25; Warwick, *Masters of Theory: Cambridge and the Rise of Mathematical Physics* (Chicago: University of Chicago Press, 2003). Warwick's research forms part of a wider movement within the history of science in which scientific meaning is reevaluated in relation to institutional, economic, and imperial interests. See also Peter Galison, *Einstein's Clocks and Poincaré's Maps: Empires of Time* (London: Sceptre, 2003), and Richard Staley, *Einstein's Generation: The Origins of the Relativity Revolution* (Chicago: University of Chicago Press, 2008). For a clear introduction to relativity, see John Stachel, "The Theory of Relativity," in *Companion to the History of Modern Science*, ed. R. C. Olby, G. N. Cantor, J. R. R. Christie, and M. J. S. Hodge (London: Routledge, 1990), 442–56. For a sociological interpretation, see Bruno Latour, "A Relativistic Account of Einstein's Relativity," *Social Studies of Science* 18 (1988): 3–44.

27. Katharine Pandora, "Popular Science in National and Transnational Perspective: Suggestions from the American Context," *Isis* 100 (2009): 356, 358.
28. Graeme Gooday, "Illuminating the Expert-Consumer Relationship in Domestic Electricity," in *Science in the Marketplace: Nineteenth-Century Sites and Experiences*, ed. Aileen Fyfe and Bernard Lightman (Chicago: University of Chicago Press, 2007), 232. See also Gooday, *Domesticating Electricity: Technology, Uncertainty and Gender, 1880–1914* (London: Pickering and Chatto, 2008).
29. Whitworth, *Einstein's Wake*, 37.
30. On the implied reader in magazines containing popular science, see Whitworth, *Einstein's Wake*, 30–31.
31. Pandora, "Popular Science," 347; Beer, *Open Fields*, 173.
32. Roger Cooter and Stephen Pumfrey, "Separate Spheres and Public Places: Reflections on the History of Science Popularization and Science in Popular Culture," *History of Science* 32 (1994): 250.
33. For a helpful overview of the debate over "popular science," see Jonathan R. Topham, "Introduction: Focus on Historicizing 'Popular Science,'" *Isis* 100 (2009): 310–18; Ralph O'Connor connects this to work on "popular culture" in "Reflections on Popular Science in Britain: Genres, Categories, and Historians," *Isis* 100 (2009): 333–45. Bowler, in *Science for All*, p. 5, discusses the development of pejorative connotations around the "popular."
34. On Eddington as Quaker astronomer and internationalist, see Matthew Stanley, *Practical Mystic: Religion, Science and A. S. Eddington* (Chicago: University of Chicago Press, 2007).
35. On Eddington's early engagement with relativity, see Stanley, *Practical Mystic* and Warwick, *Masters of Theory*.
36. For contemporaneous criticism of Eddington's style, see, for example, *Le Principe de Relativité et la Théorie de la Gravitation*, by Jean Bequerel, review by J. R., *Science Progress* 18, no. 70 (October 1923): 313–14.

37. Eddington's balancing of physics and everyday life is discussed in Beer, "Eddington and the Idiom of Modernism," 300, 302, and Beer, "'Wireless,'" 160.
38. On Sayers's uses of science, see Catherine Kenney, *The Remarkable Case of Dorothy L. Sayers* (Kent, OH: Kent State University Press, 1990), 48–49; Robert Kuhn McGregor and Ethan Lewis, *Conundrums for the Long Week-End: England, Dorothy L. Sayers, and Lord Peter Wimsey* (Kent, OH: Kent State University Press, 2000).
39. On the origins of Wimsey, see McGregor and Lewis, *Conundrums for the Long Week-End*, 20–21.
40. On the literary interest of Sayers's writing, see Kenney, *Remarkable Case*, and Gill Plain, *Women's Fiction of the Second World War: Gender, Power and Resistance* (Edinburgh: Edinburgh University Press, 1996).
41. The literary biography of Empson is *William Empson*, by John Haffenden: vol. 1, *Among the Mandarins* (Oxford: Oxford University Press, 2005), and vol. 2, *Against the Christians* (Oxford: Oxford University Press, 2006).
42. Empson's controversial engagement with Donne is discussed by John Haffenden in his introduction to William Empson, *Essays on Renaissance Literature*, vol. 1, *Donne and the New Philosophy*, ed. John Haffenden (Cambridge: Cambridge University Press, 1993).
43. The groundwork for reconnecting Empson's poetry with his diverse source materials has been established by John Haffenden, editor of William Empson, *Complete Poems* (London: Allen Lane, 2000).
44. For example, Maria DiBattista and Lucy McDiarmid, eds., *High and Low Moderns: Literature and Culture, 1889–1939* (Oxford: Oxford University Press, 1996); Patrick Collier, *Modernism on Fleet Street* (Aldershot: Ashgate, 2006); David M. Earle, *Re-covering Modernism: Pulps, Paperbacks, and the Prejudice of Form* (Farnham: Ashgate, 2009); Douglas Mao and Rebecca L. Walkowitz, eds., *Bad Modernisms* (Durham: Duke University Press, 2006); Mark Morrisson, *The Public Face of Modernism: Little Magazines, Audiences and Reception, 1905–1920* (Madison: University of Wisconsin Press, 2001); Lawrence Rainey, *Institutions of Modernism: Literary Elites and Public Cultures* (New Haven: Yale University Press, 1998).

CHAPTER ONE

1. C. Patrick Thompson, "The Einstein Year," *Time and Tide*, June 24, 1921, 599–600.
2. The reference to blue stockings derives from the eighteenth-century "Bluestocking Circle" of influential women intellectuals. Originally a reference to homely, informal blue worsted as opposed to black silk stockings, the term acquired derogatory connotations when applied to women whose intellectual aspirations made them unattractive. The women in Einstein's audience are more interested in fashion than physics, judging from their

preference for champagne stockings. If the satire had been written after the 1960s, Jack might remark that "they were supposed to burn their bras, but none of them ever did." On the Bluestocking Circle, see Elizabeth Eger and Lucy Peltz, *Brilliant Women: 18th-Century Bluestockings* (London: National Portrait Gallery, 2008).

3. C. Patrick Thompson (dates unknown) served as a lieutenant in the Royal Air Force and published a collection of air fighter stories, *Cocktails* (1919). He worked extensively for *Time and Tide*, in which his pieces were often anonymously published; see Catherine Clay, *British Women Writers 1914–1945: Professional Work and Friendship* (Aldershot: Ashgate, 2006), 72. He also contributed short fiction on adventure, mystery, and science themes to other newspapers and magazines during the 1930s and 1940s.
4. Clay, *British Women Writers*, 9. *Time and Tide* was founded and financed by Lady Margaret Rhondda. The paper did not begin to cover its expenses until 1933, by which time its founder's socialist sympathies had yielded to a more conservative vision. See *British Women Writers*, 56, and Dennis Griffiths, ed., *Encyclopedia of the British Press, 1422–1992* (Basingstoke: Macmillan, 1992), 485.
5. The Representation of the People Act, passed in June 1918, allowed women over thirty to vote. A man's right to vote was also changed from depending on property ownership to depending on a minimum period of residency. These changes in legislation are outlined in John Hostettler and Brian P. Block, *Voting in Britain: A History of the Parliamentary Franchise* (Chichester: Barry Rose, 2001), 355. The cultural context is discussed in Christopher Lawrence and Anna-K Mayer, eds., *Regenerating England: Science, Medicine and Culture in Inter-war Britain* (Amsterdam: Rodopi, 2000), 1–23.
6. Griffiths, *Encyclopedia of the British Press*, 292.
7. J. Lee Thompson, *Northcliffe: Press Baron in Politics, 1865–1922* (London: John Murray, 2000), 128.
8. The persistence of this image is explored in Peter Catterall, Colin Seymour-Ure, and Adrian Smith, eds., *Northcliffe's Legacy: Aspects of the British Popular Press, 1896–1996* (Basingstoke: Macmillan, 2000). On the "cultural narrative of journalistic crisis," see Collier, *Modernism on Fleet Street*, 7. Northcliffe's significance during the early 1900s is summarized in Matthew Kibble, "'The Betrayers of Language': Modernism and the *Daily Mail*," *Literature and History*, 3rd ser., 11, no. 1 (2002): 62–80.
9. On advertising, see Morrisson, *Public Face of Modernism*, 4. Perspectives on the "Northcliffe Revolution" are discussed in Peter Broks, *Media Science before the Great War* (Basingstoke: Macmillan, 1996), 15–16.
10. A helpful overview of these issues is given by Colin Seymour-Ure in "Northcliffe's Legacy," in *Northcliffe's Legacy*, 9–25.
11. "Light Caught Bending—A Discovery Like Newton's," *Daily Mail*, November 7, 1919, 7.
12. Letters to the Editor, *Daily Mail*, November 8, 1919, 5.

13. "Leeds in a Bad Way," *Daily Mail*, November 8, 1919, 5.
14. Philip S. Bagwell, *The Railwaymen: The History of the National Union of Railwaymen* (London: George Allen & Unwin, 1963), 344, 391.
15. Ibid., 395.
16. "Arithmetic Extraordinary," *Daily Mail*, November 8, 1919, 6.
17. "Bent Light Puzzle," 5.
18. Plot described in "The Eclipse," *London Magazine* 44, no. 113 (March 1920): 65–68.
19. On November 7, 1919, the *Daily Mail* had twelve pages and the *Times*, twenty-six. They were both broadsheets at this time.
20. "The Einstein Upheaval," *Punch*, November 26, 1919, 442.
21. "Revolution in Science—New Theory of the Universe," *Times*, November 7, 1919, 12.
22. "Labour Disputes Courts—Second Reading of the Bill," *Times*, November 7, 1919, 12.
23. I am grateful to Nick Mays of News International for providing details of authorship from marked-up archive copies of the *Times*. Peter Chalmers Mitchell (1864–1945) was a zoologist and journalist. On his science writing activities, see Bowler, *Science for All*.
24. On the involvement of physical scientists in the investigation of psychic phenomena, see Richard Noakes, "The 'world of the infinitely little': Connecting Physical and Psychical Realities circa 1900," *Studies in the History and Philosophy of Science* 39, no. 3 (2008): 323–33; Noakes, "'The Bridge which is Between Physical and Psychical Research': William Fletcher Barrett, Sensitive Flames and Spiritualism," *History of Science* 42 (2004): 419–64; Noakes, "Telegraphy Is an Occult Art: Cromwell Fleetwood Varley and the Diffusion of Electricity to the Other World," *British Journal for the History of Science* 32, no. 4 (1999): 421–59; Roger Luckhurst, *The Invention of Telepathy* (Oxford: Oxford University Press, 2002); Janet Oppenheim, *The Other World: Spiritualism and Psychical Research in England, 1850–1914* (Cambridge: Cambridge University Press, 1985).
25. "The Ether of Space—Sir Oliver Lodge's Caution," *Times*, November 8, 1919, 12.
26. "The Revolution in Science—Astronomers' Discussion," *Times*, November 15, 1919, 14.
27. E. Nevill, letter to the editor, *Times*, November 17, 1919, 8; "Sir O. Lodge on Einstein's Theory," *Times*, November 25, 1919, 16.
28. "The Revolution in Science—Reconstruction Proposed by Sir J. Larmor," *Times*, November 21, 1919, 12; Joseph Larmor, "The Einstein Theory—A Belgian Professor's Investigations," letter to the editor, *Times*, January 7, 1920, 8.
29. "Einstein on His Theory," *Times*, November 28, 1919, 13–14.
30. See Whitworth, *Einstein's Wake*, 200–201.
31. "The Fabric of the Universe," *Times*, November 7, 1919, 13.

32. "The Revolution in Science—Astronomers' Discussion," *Times*, November 15, 1919, 13. George Berkeley (1685–1753) proposed that ideas are representations of other ideas, and that the external world is an idea in the mind of God. I am grateful to Paul Dicken for clarifying this comment from the *Times*.
33. Frederic Harrison, "The Theory of Space—Practical Certainty and Relative Truth," letter to the editor, *Times*, November 21, 1919, 8.
34. Thomas Case, "Theories of Space—Newton and Einstein—The Absolute and the Relative," letter to the editor, *Times*, November 22, 1919, 8. Case's letters to the *Times* were collected and published posthumously in 1927, the cost of which he had provided for in his will.
35. "Einstein," *Times*, December 4, 1919, 15.
36. "Silence and Remembrance," *Times*, November 7, 1919, 13.
37. "Motor Show—Opening To-Day—More Cars for Women," *Times*, November 7, 1919, 13.
38. "High Milk Prices" and "Food Prices Still Higher," *Times*, November 7, 1919, 13.
39. "The Past and the Future," *Times*, November 22, 1919, 13.
40. " 'It All Depends,' " *Times*, May 27, 1921, 11.
41. Richard Burdon Haldane, Viscount, *The Reign of Relativity* (London: John Murray, 1921), 424.
42. "Astronomers on Einstein—New Geometry Wanted—Dr. Eddington and Relativity," *Times*, December 13, 1919, 9. Eddington repeated this analogy, used to explain the motion of bodies under Einstein's law of gravitation, in *The Nature of the Physical World* (Cambridge: Cambridge University Press, 1928), 126: "It is a rule of the Trade Union of matter that the longest possible time must be taken over every job."
43. " 'Which Is Absurd,' " *Daily Express*, November 8, 1919, 4. The Russian word invokes the story of Prussian ambassador Otto von Bismarck's delight in peasant philosophy, in which obstacles are "nothing," described in a letter to the *New York Times*, September 1, 1918, 27.
44. " 'Which Is Absurd,' " *Daily Express*, November 10, 1919, 6.
45. "Euclid Up-to-Date—Novel Possibilities in the 'Fourth Dimension,' " *Daily Sketch*, November 10, 1919, 2.
46. "Unpopular Science," *Daily News*, November 8, 1919, 6.
47. The relationship between the *News* and the Liberal Party is described by Stephen Koss in *Fleet Street Radical: A. G. Gardiner and the* Daily News (London: Allen Lane, 1973).
48. Anne Chisholm and Michael Davie, *Beaverbrook: A Life* (London: Hutchinson, 1992), 135.
49. Ibid., 68, 154.
50. Ibid., 148–49. Graham Moffat's comedy was first performed in London in 1910, enjoyed a year-long run on Broadway, and was made into a film in 1921.

51. Griffiths, *Encyclopedia of the British Press*, 70.
52. "Light Weighed—Greatest Discovery Since That of Gravitation," *Daily Express*, November 7, 1919, 7.
53. On trains in relativity exposition, see Whitworth, *Einstein's Wake*, 212.
54. "Cinemativity—Time and Space as Puppets of the Screen," *Daily Express*, June 14, 1921, 5.
55. "Upsetting the Universe—Dizzy Results of the New Light Discovery," *Daily Express*, November 8, 1919, 5.
56. David Prerau, *Saving the Daylight: Why We Put the Clocks Forward* (London: Granta, 2005), 53–57.
57. House of Lords debate on the Summer Time Bill, May 17, 1916. Quoted in Prerau, *Saving the Daylight*, 56.
58. Prerau, *Saving the Daylight*, 66.
59. I am grateful to Shafquat Towheed for suggesting the concept of "Eloi readers" in connection with early-twentieth-century popular science publishing during discussions at the "Publishing Science" seminar series organized by the Book History Research Group of the Open University and the Institute of English Studies, University of London, 2007–2008.
60. On contemporaneous antipathy toward *Tit-Bits*, see Jonathan Rose, "Was Capitalism Good for Victorian Literature?" *Victorian Studies* 46 (2004): 495–96; Rainey, *Institutions of Modernism*, 2; Hugh Kenner, *A Sinking Island: The Modern English Writers* (London: Barrie & Jenkins, 1988).
61. Einstein's tours of America and Europe in the spring and summer of 1921 are summarized in Friedman and Donley, *Einstein as Myth and Muse*, 17–20.
62. Strutt, *Life of Sir J. J. Thomson*, 203.
63. See introduction, note 13. The quoted speech is from Clark, *Einstein*, 267.
64. Vintage suggested by Tom Cave of Berry Brothers & Rudd Limited.
65. The Einstein dinner party guests at Viscount Haldane's residence are listed in the *Times* Court Circular for June 11, 1921, 13.
66. The *Cambridge Historical Register* records that T. G. N. (Graeme) Haldane completed part I of the Cambridge Mathematical Tripos in 1919 and part II of Natural Sciences (specializing in physics) in 1922.
67. "Einstein on His Theory—Relativity and Philosophy," *Times*, June 13, 1921, 11.
68. Frances Horner, *Time Remembered* (London: Heinemann, 1933), 9, 26; K. D. Reynolds, "Horner, Frances Jane, Lady Horner (1854/5–1940)," in *Oxford Dictionary of National Biography* (Oxford University Press, 2004), accessed April 30, 2011, http://www.oxforddnb.com/view/article/49524.
69. "Light Caught Bending," 7.
70. Thomson, *Recollections and Reflections*, 430.
71. Roy Jenkins, *Asquith*, 3rd ed. (London: Collins, 1986, first published 1964), 174, 58–59, 413.
72. Thompson, *Northcliffe*, 243; Jean Graham Hall and Douglas F. Martin,

Haldane: Statesman, Lawyer, Philosopher (Chichester: Barry Rose Law, 1996), 253–60 and 265–70.

73. Clark, in *Einstein*, 266, notes that Haldane had wanted Lloyd George to attend.
74. J. C. Segrue, "Einstein at Home," *Daily News*, June 9, 1921, 4.
75. "Relativity Chase in a Ship—How Professor Einstein Arrived," *Daily Express*, June 9, 1921, 5.
76. "Charivaria," *Punch*, June 22, 1921, 481.
77. "Einstein's Debut," *Punch*, June 22, 1921, 490.
78. "Crowds to Hear Einstein—'Relativity' Vies With Cricket," *Daily News*, June 14, 1921, 1.
79. "Music and Science," *Daily News*, June 15, 1921, 4.
80. "The Last Strawinksy," *Daily News*, June 11, 1921, 5.
81. "Some Einstein Perplexities—Danger Posts in Path of Little Knowledge," *Daily News*, June 11, 1921, 5.
82. "Crowds to Hear Einstein."
83. "Charivaria," *Punch*, June 8, 1921, 441.
84. "Why Nothing Stands Still."
85. Tim Boon, in *Films of Fact: A History of Science in Documentary Films and Television* (London: Wallflower, 2008), 7, describes the screening of *Cheese Mites*, "the sensation of the first public programme of scientific films shown at the Alhambra Music Hall in Leicester Square, London, in August 1903."
86. "Einsteinized," *Punch*, December 10, 1919, 498.

CHAPTER TWO

1. Brazier, "What Is the Use of Einstein?"
2. Noiseless Typewriter advertised in *Conquest* 1, no. 4 (February 1920).
3. [Percy Harris], "The Editor's Chair—A Short Talk on the Aims and Ideals of 'Conquest,'" *Conquest* 1, no. 1 (November 1919): 13. Information about *Conquest, Discovery*, and *Armchair Science* can be found in Peter Bowler, "Experts and Publishers: Writing Popular Science in Early Twentieth-Century Britain, Writing Popular History of Science Now," *British Journal for the History of Science* 39, no. 2 (2006): 159–87. See also Bowler, *Science for All*, chap. 9.
4. [Harris], "Editor's Chair," *Conquest* 1, no. 1 (November 1919): 13, and *Conquest* 1, no. 2 (December 1919): 71.
5. [Harris], "Editor's Chair," *Conquest* 1, no. 1 (November 1919): 13.
6. [Percy Harris], "Questions and Answers—Solutions of Readers' Difficulties," *Conquest* 2, no. 5 (March 1921): 233–34.
7. [Percy Harris], "Some Strange Questions and a Few Experiments," *Conquest* 2, no. 10 (August 1921): 421–22.
8. *Conquest* 1, no. 3 (January 1920).

9. Burrows, "Juggling with Air," 107.
10. [Harris], "Editor's Chair," *Conquest* 1, no. 3 (January 1920): 134.
11. Davidson, "Weighing Light."
12. D. N. Mallik, "Relativity of Time and Space," *Conquest* 2, no. 4 (February 1921): 182–85; J. A. Fleming, "Some Difficulties in the Theory of Relativity," *Conquest* 3, no. 10 (August 1922): 419–22.
13. For a detailed account of eclipse expeditions in relation to imperial interests, see Alex Soojung-Kim Pang, *Empire and the Sun: Victorian Solar Eclipse Expeditions* (Stanford: Stanford University Press, 2002).
14. Davidson, "Weighing Light," 130.
15. Ibid., 131.
16. Ibid., 123. Michelson himself may have been the first to use the analogy with swimmers; see Dorothy Michelson Livingston, *The Master of Light: A Biography of Albert A. Michelson* (New York: Scribner, 1973), 77. I am grateful to David Stump and Michael Fowler for drawing my attention to this source.
17. Davidson, "Weighing Light," 125.
18. [Percy Harris], "Questions and Answers," *Conquest* 1, no. 5 (March 1920): 247.
19. Whitworth, in "The Clothbound Universe," 66, discusses the possibility of qualification titles intimidating readers of popular science books, a point that may be extended to magazine articles.
20. Mallik, "Relativity of Time and Space," 182.
21. Einstein introduces the measuring rod, railway carriage, clocks, and a uniformly accelerating chest in *Relativity: The Special and the General Theory*, trans. Robert W. Lawson (London: Routledge, 1960; first published 1920 by Methuen), 5, 9, 10, and 66. On instrumentation and special relativity, see Staley, *Einstein's Generation*, chap. 3.
22. Mallik, "Relativity of Time and Space," 183.
23. Brazier, "What Is the Use of Einstein?," 123.
24. Ibid., 124. The implications of Einstein's 1905 assertion of equivalence between mass (or inertia) and energy were swiftly evoked in public discussions of general relativity in 1919 because the work of Rutherford and Soddy on matter and radiation had been well publicized during the previous ten years; see Bowler, *Science for All*, 35; Friedman and Donley, *Einstein as Myth and Muse*, chap. 6; Richard E. Sclove, "From Alchemy to Atomic War: Frederick Soddy's 'Technology Assessment' of Atomic Energy, 1900–1915," *Science, Technology and Human Values* 14, no. 2 (1989): 163–94. A semi-technical account of Einstein's work on inertia and energy is given by Max von Laue in "Inertia and Energy," in *Albert Einstein, Philosopher-Scientist*, ed. P. A. Schilpp (Evanston: Library of Living Philosophers, 1949), 501–33.
25. [Percy Harris], "Questions about Einstein," *Conquest* 3, no. 7 (May 1922): 301.
26. [Harris], "Questions and Answers," *Conquest* 3, no. 4 (February 1922): 169.

27. [Harris], "Questions about Einstein."
28. Fleming, "Some Difficulties in the Theory of Relativity," 419–20.
29. Ibid., 420.
30. "The Conquest of Dust," *Conquest* 4, no. 41 (March 1923): 175–77; "The Conquest of Drought," *Conquest* 2, no. 12 (October 1921): 515–19.
31. Bowler, *Science for All*, 163.
32. [T. Barton Kelly], "Notice to Readers," *Conquest* 7, no. 5 (March 1926): 126.
33. Bowler, *Science for All*, 173.
34. On the neglect of science debate, see Whitworth, *Einstein's Wake*, 116–20; Anna-Katherina Mayer, "Moralizing Science: The Uses of Science's Past in National Education in the 1920s," *British Journal for the History of Science* 30, no. 1 (1997): 51–52, and "Reluctant Technocrats: Science Promotion in the Neglect-of-Science Debate of 1916–1918," *History of Science* 43, no. 140 (2005): 139–59; Lawrence and Mayer, *Regenerating England*, 67–106.
35. On A. S. Russell as an antispecialist and *Discovery* as a spin-off from the neglect of science debate, see Lawrence and Mayer, *Regenerating England*, 84, 86–87.
36. Bowler, *Science for All*, 169.
37. [A. S. Russell], "Editorial Notes," *Discovery* 1, no. 1 (January 1920): 3.
38. R. S. Conway, "New Light on Old Authors I: The Secret of Philae," *Discovery* 1, no. 1 (January 1920): 4–7; Katherine A. Esdaile, "New Light on a Neglected Century of British Sculpture," *Discovery* 3, no. 33 (September 1922): 241–46; A. S. Wilson-Jones, "New Light on the Origin of Petroleum," *Discovery* 5, no. 54 (June 1924): 87; Stanley Casson, "New Light on the Ruins of Troy," *Discovery* 5, no. 49 (January 1924):14–19; A. C. Seward, "New Light on the Coal Measure Forests," *Discovery* 10, no. 113 (May 1929): 148–51. In *Science for All*, 174, Bowler observes that "even in the first issues science occupied well over half the pages, rising to 90 percent within a few years."
39. This manifesto echoes the founding editorial of *Nature*. See Beer, *Open Fields*, 257.
40. J. C. W. Reith, "The Future of Wireless Broadcasting," *Discovery* 10, no. 110 (February 1929): 37–39; O. G. S. Crawford, "The Archaeology of Tomorrow," *Discovery* 10, no. 113 (May 1929): 145–47; Colonel the Master of Semphill, "Aviation Spreads Its Wings," *Discovery* 10, no. 116 (August 1929): 247–49; Oliver Lodge, "The New Outlook in Physics," *Discovery* 10, no. 112 (April 1929): 109–12; A. C. D. Crommelin, "Astronomy Widens Its Vision," *Discovery* 10, no. 111 (March 1929): 78–80. Lodge did not mention Einstein or relativity, merely referring to "the greater generalization called space-time" (110); Crommelin's article focused on telescopes.
41. H. Spencer Jones, "Gravitation and Light," *Discovery* 1, no. 2 (February 1920): 48.
42. Harold Spencer Jones, "How Soon to the Moon?," *New Scientist* 2, no. 47 (October 10, 1957): 7–8.

43. [A. S. Russell], "Editorial Notes," *Discovery* 1, no. 3 (March 1920): 68.
44. "Books of the Month," *Discovery* 1, no 9 (September 1920): 271–72; [A. S. Russell], "Editorial Notes," *Discovery* 1, no. 10 (October 1920): 292.
45. A. S. Russell, "A Popular Exposition of Einstein's Theory," *Discovery* 3, no. 32 (August 1922): 220–21.
46. A. Vibert Douglas, "Measuring the Universe," *Discovery* 5, no. 57 (September 1924): 196–98. Vibert Douglas contributed a further six articles to *Discovery* during the 1920s. A Canadian astronomer, she had undertaken postgraduate work in contact with Eddington during the early 1920s.
47. [John A. Benn], "Editorial Notes," *Discovery* 10, no. 111 (March 1929): 71; H. F. Biggs, "The Electron and Professor Eddington," *Discovery* 10, no. 112 (April 1929): 132–34. On Eddington's fundamental theory, see Noel B. Slater, *The Development and Meaning of Eddington's "Fundamental Theory"* (Cambridge: Cambridge University Press, 1957); and Clive Kilmister, *Eddington's Search for a Fundamental Theory: A Key to the Universe* (Cambridge: Cambridge University Press, 2005).
48. On Weyl's theory, and Eddington's further elaboration, see Tom Ryckman, *The Reign of Relativity: Philosophy in Physics, 1915–1925* (New York: Oxford University Press, 2005), 77–89 and 218–234.
49. H. F. Biggs, "What the New Relativity Means," *Discovery* 10, no. 114 (June 1929): 196.
50. Biggs, "What the New Relativity Means," 197. This example is taken from the opening of Einstein's 1905 paper "On the Electrodynamics of Moving Bodies."
51. Biggs, "What the New Relativity Means," 198.
52. Ibid., 199.
53. Ibid., 200.
54. Bowler, "Experts and Publishers," 179.
55. Arnold Bennett, "The Angry Mr. Mencken: 'Lacks Balance, Sense of Justice and Evidence—But Has Done Good,'" March 31, 1927, in *Evening Standard Years*, 36.
56. Arnold Bennett, "Einstein for the Tired Business Man," April 21, 1927, in *Evening Standard Years*, 42.
57. This is also the line taken by Edwin E. Slosson in *Easy Lessons in Einstein* (London: Routledge, 1920).
58. Bennett, "Einstein for the Tired Business Man," 43.
59. Arnold Bennett, *The Night Visitor, and Other Stories* (London: Cassell, 1931), 22.
60. Bennett, "Einstein for the Tired Business Man," 44.
61. See, for example, Bennett's criticism of Laura Riding and Gertrude Stein in "The 'Monstrous Conceit' of Some Modernists," 1 March 1928, in *Evening Standard Years*, 131–33, where he also takes A. N. Whitehead to task for clumsy writing.
62. See note 34.

63. Edwin R. Nye and Mary E. Gibson, *Ronald Ross: Malariologist and Polymath* (Basingstoke: Macmillan, 1997), 210.
64. Bowler, *Science for All*, 163. There is further discussion of *Science Progress* on 167–68.
65. Ronald Ross, "Mr. Lloyd George, the Nation of Shopkeepers, and the Pied Piper of Hamelin," *Science Progress* 10, no. 38 (October 1915): 322.
66. Ibid., 318.
67. R. R. [Ronald Ross], "Mr. Man-in-the-Mass," *Science Progress* 10, no. 39 (January 1916): 485.
68. Ibid., 486.
69. Ibid., 487, 486.
70. Ross is recapitulating the Victorian assertion of a scientific elite in response to heterogeneous mass audiences discussed in Secord, *Victorian Sensation*, 46.
71. Ross's allegory may have been inspired by the writings of Ernest Renan, particularly *Caliban* (1878) and *L'avenir de la science* (1890). See Koenraad Geldof, "Look Who's Talking: Caliban in Shakespeare, Renan and Guéhenno," in *Constellation Caliban: Figurations of a Character*, ed. Nadia Lie and Theo d'Haen (Amsterdam: Rodopi, 1997), 81–112.
72. "The Principle of Relativity," *Science Progress* 9, no. 34 (October 1914): 352–53; G. W. de Tunzelman, "Physical Relativity Hypotheses Old and New," *Science Progress* 13, no. 51 (January 1919): 475–82; de Tunzelman, "The General Theory of Relativity and Einstein's Theory of Gravitation," *Science Progress* 13, no. 52 (April 1919): 652–57.
73. H. S. J. [Harold Spencer Jones], "Sir Oliver Lodge and Einstein's Theory," *Science Progress* 14, no. 55 (January 1920): 451–52; H. S. J. [Harold Spencer Jones], "Einstein on His Theory," 452.
74. D. Laugharne-Thornton, "The Einstein Theory of Relativity," *Science Progress* 16, no. 64 (April 1922): 641–43.
75. J. R. Haldane, "Synthetic Relativity," *Science Progress* 18, no. 69 (July 1923): 133–35.
76. *The Mathematical Theory of Relativity*, by A. Kopff, review by S. B., *Science Progress* 18, no. 72 (1924): 647.
77. *Le Principe de Relativité et la Théorie de la Gravitation*, by Jean Bequerel, review by J. R., *Science Progress* 18, no. 70 (October 1923): 313–14.
78. *Four Lectures on Relativity and Space*, by C. P. Steinmetz, review by J. W. F., *Science Progress* 18, no. 72 (1924): 651–52.
79. Review of *The Theory of Relativity*, by Erwin Freundlich, *Science Progress* 18, no. 72 (1924): 648.
80. Arthur Stanley Eddington, "Gravitation and the Principle of Relativity," *Nature* 101, no. 2523 (March 7, 1918): 15–17; no. 2524 (March 14, 1918): 34–36. The text of Eddington's Friday Evening Discourse from February 1, 1918, can also be found in *Proceedings of the Royal Institution* 22 (1918): 215–28.

81. The same excerpts from Russell's *ABC of Relativity* were serialized across the Atlantic during June 1925 in the *Nation*. See Friedman and Donley, *Einstein as Myth and Muse*, 14–16. The *Nation* printed two letters in July 1925 from correspondents Sydney A. Reeve of New York and C. L. E. of London, England, giving short shrift to Russell and Einstein: "Correspondence—Einstein's Relativity," *Nation* 121, no. 3133 (July 22, 1925): 116–18. Articles alongside the *ABC* included "Race Pride and Race Prejudice," "Class Justice in Germany," "International Trade-Union Unity," "Peace or 'Prosperity,'" and "A New Constitution for Russia."
82. On criticism of popular cosmology, see Bowler, *Science for All*.
83. F. M. Leventhal, *The Last Dissenter: H. N. Brailsford and His World* (Oxford: Clarendon Press, 1985), chap. 10.
84. Ibid., 176–77.
85. H. N. Brailsford, "Socialism and Property," *New Leader*, March 6, 1925, 4–5; "Industry as a Public Service," March 13, 1925, 6–7; "Who Shall Ration Work?," March 20, 1925, 4–5; "The Strategical Roads to Power," March 27, 1925, 8–9. These articles were published together in book form as *Socialism for Today* (London: Independent Labour Party, 1925). See Leventhal, *The Last Dissenter*, 189–90.
86. Yaffle, "It Occurs to Me," *New Leader*, March 27, 1925, 20.
87. Barbara Wootton, "Our Autocratic Bankers," *New Leader*, March 13, 1925, 9. Wootton (1897–1988) began her career as a Cambridge economist, moving into criminology and social work. A moving account of her life and work is given by A. H. Halsey, "Wootton, Barbara Frances, Baroness Wootton of Abinger (1897–1988)," in *Oxford Dictionary of National Biography*, (Oxford University Press, October 2009), accessed October 21, 2009, http://www.oxforddnb.com/view/article/39876.
88. Bertrand Russell, "The ABC of Relativity—1. Why Clocks and Footrules Mislead," *New Leader*, March 6, 1925, 12.
89. Bertrand Russell, "The ABC of Relativity—2. How Space and Time Are One," *New Leader*, March 13, 1925, 10.
90. Bertrand Russell, "The ABC of Relativity—3. The Eel and the Measuring Rod," *New Leader*, March 20, 1925, 12.
91. Bertrand Russell, "The ABC of Relativity—4. Nature, the Anarchist," *New Leader*, March 27, 1925, 12.
92. Canon Lewis Donaldson, "Deans as Class Warriors," *New Leader*, March 6, 1925, 3–4.
93. C. H. Douglas, "Credit-Power and Democracy—Chapter 12," *New Age* 27, no. 13 (July 29, 1920): 196.
94. The Quaker Oats advertisement is described in Aldous Huxley's novel *Point Counter Point* (Harmondsworth: Penguin, 1975; first published 1928), 298.
95. "The Romance of Rubber," *Armchair Science* 1, no. 8 (November 1929): 490–92; A. M. Low, "The Future of Women," *Armchair Science* 1, no. 12 (March 1930): 710–13; Ralph Stranger, "The Secrets of a Broadcasting Studio,"

Armchair Science 1, no. 11 (February 1930): 696–98; H. C., "The Mystery of Sleep," *Armchair Science* 1, no. 5 (August 1929): 273–75; E. N. Fox, "Learn to Fly from an Armchair," *Armchair Science* 2, no. 6 (September 1930): 341.

96. Bowler, *Science for All*, 177. See also 262.
97. "In Our Opinion," *Armchair Science* 1, no. 11 (February 1930): 646.
98. J. T. C. M.-B. [J. T. C. Moore-Brabazon], "Physics and the Ordinary Man," *Armchair Science* 2, no. 5 (August 1930): 289.
99. J. T. C. Moore-Brabazon, "Some Thoughts on Current Physics," *Armchair Science* 2, no. 11 (February 1931): 622.
100. Moore-Brabazon, "Some Thoughts on Current Physics," 623.
101. Ibid., 622.
102. J. W. Dunne, "Can We Travel in the Time Dimension?," *Armchair Science* 3, no. 7 (October 1931): 415–16. Moore-Brabazon was a member of the Royal Aero Club and a pioneering aviator, and Dunne was a prominent aeronautical engineer. They probably knew each other long before the publication of *An Experiment with Time* (1927).
103. "Relativity," *Armchair Science* 1, no. 11 (February 1930): 652.
104. Bowler, in *Science for All*, 180, notes that from the mid-1930s *Armchair Science* made "extensive use of syndicated material, much of it from America, so the magazine took on the character of a *Tit-Bits* for the sciences, in effect replacing *Popular Science Siftings*, which had closed down in 1927."
105. "Relativity," 654.
106. A. M. Low, "There's Relativity in Everything," *Armchair Science* 3, no. 11 (February 1932): 646.
107. Ibid., 647.
108. A. M. Low, "Can Science Explain Love?," *Armchair Science* 1, no. 11 (February 1930): 648–51.
109. [Moore-Brabazon], "Physics and the Ordinary Man," 289. On Victorian debates over the flux of knowledge as vitality or failure, see Herbert, *Victorian Relativity*, 129.
110. Thomson, *Recollections and Reflections*, 430–31.

CHAPTER THREE

1. [Harris], "Questions about Einstein," 301.
2. Mike Ashley, *The Age of the Storytellers: British Popular Fiction Magazines 1880–1950* (London: British Library, 2006), s.vv. *Red Magazine*, 175–79. Ashley is an invaluable resource, giving details of ownership, circulation, cover price, readership, editors, contributors, and content for all the magazines discussed in this chapter.
3. The titles of the *Red* and its companion magazine, the *Yellow*, refer simply to color themes for cover art. Ashley, in *Age of the Storytellers*, 178, notes that two further magazines, *The Violet* (for light romance) and *The Green* (for sporting and outdoor tales), began in July and November 1922, respectively.

4. H. G. Wells's *The Time Machine* (1895) is an obvious precedent. "The Plattner Story" (1896) also recounts a fourth-dimensional adventure, while *The Invisible Man* (1897) and *The Wonderful Visit* (1895) make reference to four dimensions. Mike Ashley, in *Time Machines: The Story of the Science Fiction Pulp Magazines from the Beginning to 1950* (Liverpool: Liverpool University Press, 2000), 11, refers to faster-than-light travel in "The Last Days of Earth" by George C. Wallis (1901). Stories with alternative time frames were already in circulation at the time of the initial Einstein sensation; see, for example, "Mr. Brundell in Utopia," by W. L. George, in *Novel Magazine* (December 1919). *Lumen* (1872), by French astronomer Camille Flammarion (1842–1925), is the classic source for an observer traveling faster than light and witnessing past events taking place on the earth; Flammarion's significance is discussed in Whitworth, *Einstein's Wake*, 174–75. Slosson's *Easy Lessons in Einstein* includes a helpful summary of examples from fiction and popular science, including George Macdonald's *Lilith* (1895) and Wells's "The Story of Davidson's Eyes" (1895).
5. What little is known about Coutts Brisbane (real name Robert Coutts Armour, who also wrote popular fiction as Reid Whitly or Whitley) is given in Ashley, *Time Machines*, 11–12, 131, and Ashley, *Age of the Storytellers*, 176–77, 237–38. A brief entry in Paul Collins, ed., *MUP Encyclopaedia of Australian Science Fiction and Fantasy* (Carlton, Vic.: Melbourne University Press, 1998), describes him as "something of a recluse."
6. Coutts Brisbane, "Mr. Fipkins and the Younger World," *Red Magazine* 57, no. 348 (October 12, 1923): 77.
7. Ibid., 78.
8. Ibid., 78–79.
9. Ibid., 79.
10. Ibid., 84.
11. Ibid., 82.
12. Ibid., 85.
13. It is possible that Brisbane may have Einstein's finite universe in mind here, though the association is not essential for the plot. The idea of a world closed off in four-dimensional curvature, with "nowhere" outside it, was discussed by Arthur Eddington in 1926 and used by William Empson in 1928–1935 (see chap. 6).
14. Brisbane, "Mr. Fipkins," 82.
15. Ibid., 80.
16. Ibid., 85.
17. A helpful introduction to this theme can be found in Steven Matthews, *Modernism* (London: Arnold, 2004), chap. 4. James Frazer's pre-World War I warnings about the eruption of underlying savagery are discussed in Herbert, *Victorian Relativity*, 203–8.
18. Derek Edgell, in *The Order of Woodcraft Chivalry, 1916–1949, as a New Age*

Alternative to the Boy Scouts (Lewiston: Mellen, 1992) discusses one such movement in detail.

19. Ashley, *Age of the Storytellers*, 175.
20. Ibid., 175, 179.
21. Sieveking (1896–1972) was a pioneer of radio drama and subsequently worked as a television producer.
22. L. de G. Sieveking, "The Lost Omnibus," *Hutchinson's Mystery-Story Magazine* 3 (April 1924): 3.
23. Ibid., 4.
24. On subject position as created by marginal detail, see Whitworth, *Einstein's Wake*, 30.
25. *Yellow Magazine* 16, no. 101 (July 24, 1925): 331.
26. *London Magazine* 46, no. 126 (April 1921): xiv.
27. *Red Magazine* 40, no. 334 (March 30, 1923): 333.
28. A. E. Ashford, "The Time-Adjuster," *Red Magazine* 40, no. 334 (March 30, 1923): 328.
29. Ibid., 325.
30. On commercial contexts for Einstein themes in America, see Steinman, *Made in America*, and White, "Advertising Localist Modernism."
31. Ashford, "Time-Adjuster," 327.
32. Coutts Brisbane, "Thus Said Pel!," *Red Magazine* 44, no. 263 (July 9, 1920): 210.
33. Ibid., 211.
34. Ibid., 209.
35. Ibid., 212.
36. Leslie Beresford, "The Stranger from Somewhere," *Red Magazine* 56, no. 345 (August 31, 1923): 476–77.
37. "Breaking Up the Atom: Professor Rutherford's Discoveries," *Daily Mail*, June 11, 1921, 5.
38. Beresford, "Stranger from Somewhere," 478.
39. Ibid., 478–79. Eddington, in *Space Time and Gravitation* (Cambridge: Cambridge University Press, 1920), 26–27, describes the "favourite device" of expositors, a "voyage through space with the velocity of light" after which the traveler finds himself centuries younger than those who have stayed behind on earth. The origin of this device, also known as the "twins paradox," is generally ascribed to Paul Langevin, in "L'evolution de l'espace et du temps," *Scientia* 10 (1911): 31–54.
40. Beresford, "Stranger from Somewhere," 479.
41. Ibid., 483.
42. Ibid., 479.
43. Ibid., 482.
44. Ibid., 484.
45. Ibid., 483.

46. Ibid., 485.
47. Ashley, *Time Machines*, 12.
48. Ashley, *Age of the Storytellers*, 178–79.
49. Coutts Brisbane, "Ex Terra (Special!)" [moon attack], *Yellow Magazine* 2, no. 12 (February 24, 1922); "The Almighty Atom," *Red Magazine* 50, no. 307 (March 17, 1922); "The End—and the Beginning" [adventure on Mars], part 1, *Yellow Magazine* 15, no. 99 (June 26, 1925); part 2, *Yellow Magazine* 16, no. 100 (July 10, 1925); "Not an Earthly" [on Saturn], *Yellow Magazine* 9, no. 54 (October 5, 1923); "Under the Moons" [on Saturn], *Red Magazine* 40, no. 237 (July 11, 1919); "Where It Was Dark" [on Mercury], *Yellow Magazine* 1, no. 7 (December 16, 1921); "The Xtraordinary Mrs Grady" [shrinking elixir], *Yellow Magazine* 13, no. 84 (November 28, 1924); "High Tensions" [cyclone machine], *Yellow Magazine* 13, no. 81 (October 17, 1924); "Light Clothing," *Yellow Magazine* 18, no. 117 (March 5, 1926); "For the Good of Creation" [radiology and short skirts], *Yellow Magazine* 19, no. 120 (April 16, 1926).
50. Coutts Brisbane, "An Elementary Affair," *Red Magazine* 47, no. 350 (November 9, 1923): 260.
51. Ibid., 261, 264, 265.
52. Paul Brians, in *Nuclear Holocausts: Atomic War in Fiction* (Kent: Kent State University Press, 1987), traces early stories of destructive atomic power from 1895 onward, citing Soddy as an influential popularizer. See also Jeff Hughes, "Radioactivity and Nuclear Physics," in *Cambridge History of Science*, vol. 5, *The Modern Physical and Mathematical Sciences*, ed. Mary Jo Nye (Cambridge: Cambridge University Press, 2003), 355; Sclove, "From Alchemy to Atomic War," 177–79; and for a more extensive discussion, Friedman and Donley, *Einstein as Myth and Muse*, chap. 6.
53. Coutts Brisbane, "The Almighty Atom," *Red Magazine* 50, no. 307 (March 17, 1922): 614.
54. T. C. Wignall and G. D. Knox, "Atoms," *Yellow Magazine* 5, nos. 29–34 (October 20–December 29, 1922).
55. Ronald Ross, "A Scientific Romance," *Science Progress* 18, no. 70 (October 1923): 279.
56. A. E. Ashford, "The Q-Ray," *Red Magazine* 44, no. 263 (July 9, 1920): 195.
57. Ashford, "Q-Ray," 197. The Derby as a classic test features in a short film with visions of future London, *A Fugitive Futurist: A Q-Riosity by Q* (1924).
58. Ashford, "Q-Ray," 198.
59. Ibid., 201.
60. Ibid., 201–2.
61. Ibid., 202.
62. The uncanny qualities of wireless broadcasting are discussed in Beer, "'Wireless,'" in relation to Kipling's short story "Wireless" (1903).
63. "Cinemativity," 5; Low, "There's Relativity in Everything," 647.
64. On themes of time and space in technology and culture during the decades

before 1919, see Stephen Kern, *The Culture of Time and Space: 1880–1918* (Cambridge, MA: Harvard University Press, 1983).

65. On early broadcasting and the war, see Asa Briggs, *The History of Broadcasting in the United Kingdom*, vol. 1, *The Birth of Broadcasting* (Oxford: Oxford University Press, 1995).
66. The extensive ironies of the Great War and their lasting significance are explored in the classic work by Paul Fussell, *The Great War and Modern Memory* (Oxford: Oxford University Press, 1975). On propaganda, see Tate, *Modernism, History and the First World War*, chap. 2.
67. "Wanted: A Popular Science Picture Theatre," *Conquest* 3, no. 10 (August 1922): 423.
68. Frank Hurley's footage of the 1914–1916 Antarctic expedition was released as *South: Sir Ernest Shackleton's Glorious Epic of the Antarctic* in 1919.
69. Todd Avery, *Radio Modernism: Literature, Politics and the BBC, 1922–1938* (Aldershot: Ashgate, 2006), 7.
70. On radio's associations with the uncanny and with the problems of highbrowism and mass audience, see Beer, "Eddington and the Idiom of Modernism."
71. For example, the American mass-market author H. P. Lovecraft, whose representation of space-time owes something to his reading of Eddington. See Introduction, note 18, for works discussing the spatial fourth dimension as distinct from Einsteinian space-time. See also Clarke, *Energy Forms*, on the spatial fourth dimension and space-time in literature.
72. In his entry on the *Novel Magazine*, Ashley, in *Age of the Storytellers*, 141, notes that its name indicated "novelty" rather than the serialization of novels.
73. Coutts Brisbane's story "The End—and the Beginning," discussed in chapter 6, is an example. Other wireless stories from the *Yellow* during the early 1920s include Owen Oliver, "A Martyr to Wireless," 11, no. 68 (April 18, 1924); and Alan J. Thompson, "A Call from 2LO," 16, no. 100 (July 10, 1925).
74. E. W. Morrison, "A Wireless Uncle," *Novel Magazine* 39 (May 1924): 117–20.
75. H. W. Leggett, "Whispers on the Wind," *Novel Magazine* 37 (August 1923): 416.
76. Ibid., 418. The astronomy book may well be *The Splendour of the Heavens*, published by Hutchinsons and issued in twenty-four fortnightly parts from May 14, 1923, to March 27, 1924, each issue costing 1s. 3d.
77. E. W. Morrison, "The Bell That Never Rang," *Novel Magazine* 38 (February 1924): 411.
78. Ibid., 414.
79. Lee Foster Hartman, "The Altar of Destiny," *Novel Magazine* 37 (August 1923): 408–15. Foster Hartman went on to become editor of *Harper's Magazine* from 1931 until his death in 1941.
80. R. L. Dearden, "The Morning Star," *Novel Magazine* 40 (January 1925): 315.
81. Ibid., 316.

82. Ibid., 316, 318.
83. Ibid., 316.
84. Ibid., 317.
85. Ibid., 318.
86. On women wranglers, see Warwick, *Masters of Theory*, 281, 218.
87. These developments are summarized in Iwan Rhys Morus, *When Physics Became King* (Chicago: University of Chicago Press, 2005), 194.
88. Ibid., 193, 202–9.
89. Ashley, *Age of the Storytellers*, 142–43.
90. On Ertz, see Grant M. Overton, *Women Who Make Our Novels* (Freeport, NY: Books for Libraries, 1967), 113–19.
91. Susan Ertz, "Relativity and Major Rooke," *Harper's Magazine* 148 (April 1924): 672.
92. Ibid., 674–75.
93. William James, *Varieties of Religious Experience: A Study in Human Nature* (London: Routledge, 2002; first published 1902), 385. Emphasis in original. Quoted in Ertz, "Relativity and Major Rooke," 675.
94. Ertz, "Relativity and Major Rooke," 676.
95. Leggett, "Whispers on the Wind," 416.
96. *Punch*, December 11, 1929, 670.
97. *Punch*, June 11, 1930, 654.
98. Ashley, *Age of the Storytellers*, 196–208.
99. The fourth dimension is also mentioned at the start of Wodehouse's "The Story of Webster," in *Mulliner Nights* (London: Herbert Jenkins, 1933).
100. P. G. Wodehouse, "The Amazing Hat Mystery," *Strand Magazine*, June 1934, 563.
101. Ibid., 573.
102. P. G. Wodehouse, "Archibald and the Masses," *Strand Magazine*, February 1936, 386–97.
103. Ibid., 387.
104. Dorothy L. Sayers, "Absolutely Elsewhere," *Strand Magazine*, February 1934, 185.
105. Beer, *Open Fields*, 178, comments on the "condensed lucidity" of Eddington, acknowledging that "the burst of clarity is not secure for ever."

CHAPTER FOUR

1. Eddington's radio talks include "Matter in Interstellar Space" (April 1929), "Eclipses of the Sun" (June 1937), "Other Worlds" (December 1940), "The Spirit of Science" (February 1941), and "Your Questions Answered" (March 1944).
2. [H. H. Turner], "From an Oxford Note-Book," *Observatory* 42, no. 546 (December 1919): 454. Stanley identifies the author of this regular column in *Practical Mystic*, 86.

3. M. G. Boccardi, "German Science and Latin Science," *Observatory* no. 504 (September 1916): 381–91.
4. A. S. Eddington, "The Future of International Science," *Observatory* no. 501 (June 1916): 271.
5. [H. H. Turner], "From an Oxford Note-Book," *Observatory* no. 502 (July 1916): 323.
6. Stanley, *Practical Mystic*, chap. 3.
7. Historians and scientists have debated whether Eddington was biased in his treatment of the eclipse results. The debates are reviewed in Stanley, *Practical Mystic*, 122–23, 268–69, and Alistair Sponsel, "Constructing a 'Revolution in Science': The Campaign to Promote a Favourable Reception for the 1919 Solar Eclipse Expeditions," *British Journal for the History of Science* 35, no. 4 (2002): 439–67.
8. Cooter and Pumfrey, "Separate Spheres and Public Places," 250.
9. Close relations between British eclipse expedition planners and the *Times* had been established during the late nineteenth century, when the formation of a Joint Permanent Eclipse Committee (JPEC) of the RS and RAS helped to secure funding for these expensive expeditions and ensure a continued public profile for astronomy. See Pang, *Empire and the Sun*, 33–39. Sponsel discusses public relations work by astronomers around the 1919 eclipse announcements in "Constructing a 'Revolution in Science.'"
10. Arthur Eddington, "Einstein's Theory of Space and Time," *Contemporary Review* 116, no. 648 (December 1919): 639–43; Eddington, "Einstein on Time and Space," *Quarterly Review* no. 462 (January 1920): 226–36.
11. On Eddington's storytelling as epistemology, see Beer, "Eddington and the Idiom of Modernism."
12. I have not found a source for the comment sometimes cited on the Internet that Einstein thought Eddington's Romanes Lecture a good explanation of his theory.
13. Eddington, *Nature of the Physical World*, 289. Russell's "about-face with respect to Eddington's capabilities" is described in Stanley, *Practical Mystic*, 215–19.
14. Arthur Eddington, *The Theory of Relativity and Its Influence on Scientific Thought* (Oxford: Clarendon, 1922), 9–10.
15. Eddington, *Nature of the Physical World*, 11.
16. Beer, "Eddington and the Idiom of Modernism," 301–2.
17. Eddington, *Theory of Relativity*, 10–13.
18. Eddington, *Nature of the Physical World*, 9–10.
19. Ibid., 14.
20. Russell, *ABC of Relativity*, 219.
21. Ibid., 219–20.
22. Ibid., 15.
23. Eddington's Notebook, Add. Ms. b. 48, Eddington Papers, Wren Library, Trinity College, University of Cambridge.

24. Eddington, *Theory of Relativity*, 20.
25. Ibid., 25.
26. Eddington, *Nature of the Physical World*, 115.
27. Ibid., 116.
28. Ibid., 117.
29. Fleming, "Some Difficulties in the Theory of Relativity."
30. Eddington, *Nature of the Physical World*, 59.
31. Beer, "Eddington and the Idiom of Modernism," 299.
32. [Moore-Brabazon], "Physics and the Ordinary Man," 289–90.
33. Victor Pullin, review of *The Nature of the Physical World*, by Arthur Eddington, *Discovery* 10, no. 111 (March 1929): 104.
34. On criticism of Eddington and Jeans, see Bowler, *Reconciling Science and Religion* and *Science for All*, and Whitworth, "The Clothbound Universe" and *Einstein's Wake*.
35. Herbert Samuel, "Cause, Effect, and Professor Eddington," *Nineteenth Century and After* 113 (April 1933): 471.
36. Whitworth, "The Clothbound Universe," 72.
37. L. S. Stebbing, *Philosophy and the Physicists* (Harmondsworth: Penguin, 1944; first published 1937), 5.
38. Stebbing, *Philosophy and the Physicists*, 13.
39. "The Alarming Astronomers" is the title of part 1 of Stebbing's book; these comments are made on pp. 22–23, at the start of a section titled "The Escape of Sir James Jeans," in a general statement that includes Eddington.
40. Eddington, *Nature of the Physical World*, 37.
41. Ibid., 37–38.
42. Ibid., 42–43. The use of a worm to denote the human inhabitant of Einstein's universe had been given an amusing turn in Slosson, *Easy Lessons in Einstein*.
43. Eddington, *Nature of the Physical World*, 48.
44. Ibid., 50.
45. Ibid., 49.
46. Marilyn Bailey Ogilvie, "Obligatory Amateurs: Annie Maunder (1868–1947) and British Women Astronomers at the Dawn of Professional Astronomy," *British Journal for the History of Science* 33, no. 1 (2000): 74–76. On Walter Maunder and the BAA, see Bowler, *Science for All*, 180–81.
47. The involvement of women astronomers in the BAA is discussed in Bailey Ogilvie, "Obligatory Amateurs," 77–79.
48. Report of Meeting November 1918, *Journal of the British Astronomical Association* 29, no. 2 (November 1918): 35.
49. Ibid., 36.
50. Alan Sandage, *Centennial History of the Carnegie Institution of Washington*, vol. 1, *The Mount Wilson Observatory* (Cambridge: Cambridge University Press, 2004), chap. 26.
51. Report of Meeting November 1918, 36.

52. Ibid., 36–37.
53. Ibid., 37.
54. Ibid., 38.
55. Distances calculated by Eddington in a letter to his mother: A. S. Eddington to Sarah Ann Eddington, April 20, 1919, A4/6, Eddington Papers.
56. A. S. Eddington to Winifred Eddington, May 5, 1919, A4/8, Eddington Papers.
57. A. S. Eddington to Sarah Ann Eddington, April 13, 1919, A4/5, Eddington Papers.
58. A. S. Eddington to Sarah Ann Eddington, April 29, 1919, A4/7, and March 27, 1919, A4/3, Eddington Papers.
59. A. S. Eddington to Sarah Ann Eddington, April 29, 1919, and June 21, 1919, A4/9, Eddington Papers.
60. A. S. Eddington to Sarah Ann Eddington, April 20, April 29, and June 21, 1919, Eddington Papers.
61. See note 7.
62. Eddington, *Nature of the Physical World*, 270. A similar analogy, comparing the abstraction of physical law to a telephone exchange, was used by Eddington in "The Domain of Physical Science," in *Science, Religion, and Reality*, ed. Joseph Needham (London: Sheldon, 1925), 202.
63. Eddington's Notebook, Add. Ms. b. 48, Eddington Papers.
64. The friendship between Eddington and Trimble is described in A. Vibert Douglas, *The Life of Arthur Stanley Eddington* (London: Nelson, 1956), 7, 32–35, 183, and in Stanley, *Practical Mystic*, 33, 252.
65. Vibert Douglas, *Life of Arthur Stanley Eddington*, 34.
66. A. S. Eddington to Sarah Ann Eddington, April 13, 1919; A. S. Eddington to Winifred Eddington, May 5, 1919, Eddington Papers.
67. A. S. Eddington to Winifred Eddington, May 5, 1919, Eddington Papers.
68. A. S. Eddington to Sarah Ann Eddington, April 6, 1919, A4/4, Eddington Papers.
69. Stanley, in *Practical Mystic*, 252, comments on an unsuccessful attempt to make something of the "thin circumstantial evidence" for Eddington's homosexuality by Arthur Miller in *Empire of the Stars: Friendship, Obsession and Betrayal in the Quest for Black Holes* (London: Little, Brown, 2005).
70. Eddington, *Nature of the Physical World*, 119. On Eddington's interpretation of curvature in relativity theory in relation to his idealist philosophy, see Ryckman, *Reign of Relativity*, 209–13.
71. Ibid., 120.
72. Ibid., 139.
73. Ibid., 141.
74. Ibid., 143.
75. Stebbing, *Philosophy and the Physicists*, 22.
76. Eddington, *Nature of the Physical World*, 143.
77. For discussion of Eddington's restriction of physical law, see Loren Graham,

Between Science and Values (New York: Columbia University Press, 1981), and Bertrand Russell, *The Scientific Outlook* (London: George Allen & Unwin, 1931).

78. Eddington, *Nature of the Physical World*, 260–65. On Eddington's cyclic method, see Beer, "Eddington and the Idiom of Modernism," 301.
79. Eddington contributed two articles on relativity to the journal *Mind* in 1920: "The Meaning of Matter and the Laws of Nature According to the Theory of Relativity," *Mind* 29 (1920): 145–58, and "The Philosophical Aspect of the Theory of Relativity," *Mind* 29 (1920): 415–22.
80. Eddington, *Nature of the Physical World*, 241.
81. Ibid., xv.
82. Ibid., xvi.
83. Victorian precedents for Eddington's approach to symbolism are discussed in Herbert, *Victorian Relativity*, 184. Analogies with reading were a staple of Victorian popular science: see Lightman, *Victorian Popularizers of Science*, 189, 455. A more recent example appears in Matt Ridley's *Genome: The Autobiography of a Species in 23 Chapters* (London: Fourth Estate, 1999), 6–8.
84. Eddington, *Nature of the Physical World*, xvi–xvii.
85. Ross, "Mr. Man-in-the-Mass."
86. Eddington, *Nature of the Physical World*, 288–89.
87. [Moore-Brabazon], "Physics and the Ordinary Man," 289–90.
88. Manuscript 0.11.25, Eddington Papers.
89. Literary authors who responded to Eddington include John Steinbeck, Georges Bataille, Virginia Woolf, William Empson, Mary Butts, Robert Frost, Dorothy L. Sayers, and Aldous Huxley. For Bataille, see Parkinson, *Surrealism, Art, and Modern Science*. For Woolf, see Beer, "Eddington and the Idiom of Modernism" and *Wave, Atom, Dinosaur*. For Empson, see John Haffenden's notes in Empson, *Complete Poems*, and my two essays "Flame far too hot: William Empson's Non-Euclidean predicament," *Interdisciplinary Science Reviews* 30, no. 4 (2005): 312–22, and "Monogamy and the Next Step? William Empson and the Future of Love in Einstein's Universe," in *Some Versions of Empson*, ed. Matthew Bevis (Oxford: Oxford University Press, 2007), 242–63. For Butts, see *The Journals of Mary Butts*, ed. Nathalie Blondel (New Haven: Yale University Press, 2002), 356–57, 366. For Frost, see Guy Rotella, "Comparing Conceptions: Frost and Eddington, Heisenberg and Bohr," *American Literature* 59 (1987): 167–90. For Sayers, see my essay "'On the Back of the Light Waves': Novel Possibilities in the Fourth Dimension," in *Literature and Science*, ed. Sharon Ruston (Cambridge: Brewer, 2008), 91–110.
90. [John Middleton Murry], "The Mysticism of Modern Science," *New Adelphi* 2, no. 4 (June–August 1929): 292.
91. Geoffrey Sainsbury, "The Nature of the Physical World," *New Adelphi* 2, no. 4 (June–August 1929): 357.

92. John Macmurray, review of *The Nature of the Physical World*, by Arthur Eddington, *Criterion* 8, no. 33 (July 1929): 706.
93. T. S. Eliot, "Religion and Science: A Phantom Dilemma," *Listener*, March 23, 1932, 429.
94. Stanley, *Practical Mystic*, 223.
95. Ibid., 224. See also Stanley's essay "Mysticism and Marxism: A. S. Eddington, Chapman Cohen, and Political Engagement Through Science Popularization," *Minerva: A Review of Science, Learning and Policy* 46, no. 2 (2008): 181–94, and Bowler, *Science for All*.
96. Huxley, *Point Counter Point*, 65.
97. Ibid., 158.
98. On the impact of Mach and literary discussion of the view that physical law is a convention of description, see Whitworth, *Einstein's Wake*, chap. 3.
99. Auden and MacNeice, *Letters from Iceland*, 103.
100. Stephen Dorril, *Blackshirt: Sir Oswald Mosley and British Fascism* (London: Viking, 2006), 390–97.

CHAPTER FIVE

1. Alice Crawford, *Paradise Pursued: The Novels of Rose Macaulay* (Madison, NJ: Fairleigh Dickinson University Press 1995), 22.
2. Ibid., 64. On *Potterism*, see also Collier, *Modernism on Fleet Street*, 146–53.
3. For a friendly introduction to free indirect style, see John Mullan, *How Novels Work* (Oxford: Oxford University Press, 2006), 76, and Stevenson, *Modernist Fiction*, 32.
4. Rose Macaulay, *Potterism: A Tragi-Farcical Tract* (Brussels: Collins, 1921; first published 1920).
5. Ibid., 55, 11, 86, 124.
6. Ibid., 30.
7. Ibid., 55.
8. Ibid., 230.
9. In the redrawing of European boundaries after World War I, Italy claimed possession of Fiume, today known as the Croatian seaport Rijeka.
10. Macaulay, *Potterism*, 230–31.
11. Ibid., 231.
12. Ibid., 232.
13. Crawford, in *Paradise Pursued*, 58–59 and 62–64, describes the "clashing relativities" of *Potterism* as a dramatization of Einstein's theory in the form of a detective fiction, observing that William James's theory of personality and the writings of C. S. Peirce and Henri Bergson are also relevant influences.
14. Crawford, *Paradise Pursued*, 64.
15. Crawford, in *Paradise Pursued*, 58, discusses Macaulay's philosophical interests.

16. Whitworth, *Einstein's Wake*, 213.
17. Whitworth, " 'Pièces d'identité,' " 149.
18. Whitworth, *Einstein's Wake*, 37–38.
19. Bradshaw, "Best of Companions [Part 1]," 192; Whitworth, " 'Pièces d'identité,' " 152.
20. Bradshaw, "Best of Companions [Part 1]," 196.
21. Bradshaw, "Best of Companions [Part 1]," 197; Aldous Huxley, *Crome Yellow* (London: Granada, 1982; first published 1921), 19.
22. Whitworth, *Einstein's Wake*, 211, 214.
23. Huxley, *Crome Yellow*, 6.
24. Whitworth, *Einstein's Wake*, 212.
25. Huxley, *Crome Yellow*, 50–51.
26. Ibid., 52.
27. Ibid., 51.
28. Ibid., 10.
29. On Sayers's exploration of marriage, see Susan J. Leonardi, *Dangerous by Degrees: Women at Oxford and the Somerville College Novelists* (New Brunswick: Rutgers University Press, 1989), chap. 4.
30. A manuscript of *The Documents in the Case*, held at the Marion E. Wade Center, Wheaton College, reveals how Sayers worked references to Einstein and Eddington into the texture of the characters' interactions.
31. The extent of Barton's contribution to *The Documents in the Case* is discussed by Kenney in *Remarkable Case*, 48–49, 142.
32. Kenney, *Remarkable Case*, 142.
33. Ibid., 141.
34. Ibid., 142.
35. Dorothy L. Sayers, *The Documents in the Case* (London: New English Library, 1973; first published 1930), 12.
36. Ibid., 13.
37. Ibid., 26.
38. Ibid., 27.
39. Ibid., 62.
40. Ibid., 24. Contradictions between Victorian progressive beliefs and the findings of contemporary physics are discussed in Beer, *Open Fields*, 219.
41. Sayers, *Documents in the Case*, 24–25.
42. Ibid., 25.
43. *The Letters of Dorothy L. Sayers*, ed. Barbara Reynolds (Cambridge: Dorothy L. Sayers Society, 1998), 338.
44. Sayers, *Documents in the Case*, 25.
45. Ibid., 66.
46. Ibid., 67.
47. Bennett, *Evening Standard Years*, 44.
48. Sayers, *Documents in the Case*, 50.
49. Ibid., 50–51.

50. Ibid., 51.
51. Ibid., 67.
52. Ibid., 64. The novel's title is given on 62. The list of topics echoes that given by Rose Macaulay in a 1921 essay on how newspapers write about women, "Woman: The Eternal Topic." See Collier, *Modernism on Fleet Street*, 159–60.
53. T. S. Eliot, *Complete Poems and Plays* (London: Faber, 1969), 39. The ambivalent engagement of Eliot and other canonical modernist authors with the new physics is thoroughly explored in Whitworth, *Einstein's Wake*, with briefer, introductory accounts in Whitworth, "Physics: 'A strange footprint,'" in *A Concise Companion to Modernism*, ed. David Bradshaw (Oxford: Blackwell, 2003), 200–220, and Whitworth, "The Physical Sciences," in *A Companion to Modernist Literature and Culture*, ed. David Bradshaw and Kevin J. H. Dettmar (Oxford: Blackwell, 2006), 39–49.
54. T. S. Eliot, "Ben Jonson," cited in Whitworth, *Einstein's Wake*, 215.
55. Whitworth, in *Einstein's Wake*, 214–20, discusses the Ben Jonson essay and Eliot's early knowledge of the new geometry.
56. Sayers, *Documents in the Case*, 27.
57. Ibid., 35.
58. Eliot, "Religion and Science," 429.
59. Victorian debates over materialism, including questions about the origin of life and the new physics of the early twentieth century, are summarized in Frederick Gregory, "Intersections of Physical Sciences and Western Religion in the Nineteenth and Twentieth Centuries," in *Cambridge History of Science*, vol. 5, 43–48. In a "celebrated" lecture, "Stereochemistry and Vitalism," at the British Association in 1898, F. R. Japp drew vitalist conclusions from the asymmetry of organic compounds (by which they may be distinguished from synthetic counterparts). See D'Arcy Wentworth Thompson, *On Growth and Form*, abr. and ed. John Tyler Bonner (Cambridge: Cambridge University Press, 1992; first published 1917), 138, and Paolo Palladino, "Stereochemistry and the Nature of Life: Mechanist, Vitalist and Evolutionary Perspectives," *Isis* 81 (1990): 44–67.
60. Sayers, *Documents in the Case*, 35.
61. Ibid., 51.
62. The importance of such circles in providing access to science themes and a context for relating them to literary endeavour is discussed thoroughly in Whitworth, "'Pièces d'identité'" and *Einstein's Wake*. On Woolf and popular physics, see Beer, "Eddington and the Idiom of Modernism," "'Wireless,'" *Open Fields*, and *Wave, Atom, Dinosaur*; Banfield, *Phantom Table*, and Henry, *Virginia Woolf*.
63. Dorothy Sayers, *Further Papers on Dante* (London: Methuen, 1957), 78.
64. Eddington, *Nature of the Physical World*, xii, quoted in Sayers, *Further Papers on Dante*, 80.
65. Sayers, *Further Papers on Dante*, 80–81.

66. Ibid., 88.
67. Ibid., 85–86.
68. Ibid., 85.
69. Ibid., 86.
70. Beer, "Eddington and the Idiom of Modernism," 299.
71. Plain, *Women's Fiction*, 51.
72. On Sayers's depiction of a corresponding all-female environment from which Harriet Vane emerges to pursue a life's work with Wimsey, see Leonardi, *Dangerous by Degrees*, chap. 3.
73. Kenney, *Remarkable Case*, 48.
74. Dorothy L. Sayers, *Hangman's Holiday* (London: New English Library, 1982; first published 1933), 11.
75. Ibid., 11, 20.
76. Ibid., 12.
77. Ibid., 11.
78. Ibid., 12.
79. Ibid., 13.
80. Ibid., 15.
81. Ibid., 18.
82. Ibid., 19–20.
83. Plain, *Women's Fiction*, 51.
84. Sayers, *Hangman's Holiday*, 12.
85. Ibid., 13.
86. Barbara Reynolds, *Dorothy L. Sayers: Her Life and Soul* (London: Hodder & Stoughton, 1993), chaps. 10 and 11.
87. Sayers, *Hangman's Holiday*, 30.
88. Sayers, "Absolutely Elsewhere," 185.
89. This connotation emerged strongly during a debate over BBC radio content during the early 1930s, discussed in Beer, "'Wireless,'" 164.
90. Dorothy L. Sayers, "Impossible Alibi," *Mystery: The Illustrated Detective Magazine*, January 1934, 19. Eddington's name was certainly not unfamiliar in the United States, as coverage in the *New York Times* indicates ("How Tall Are You, Einstein Measure? Prof. Eddington, 6 Feet to the Eye, Explains How It May Be Really Only 3 Feet," December 4, 1919, 19). The *Mystery* editor may have recognized that it didn't carry the same sense of social grading for American audiences as it did in Britain.
91. Nicola Humble, *The Feminine Middlebrow Novel, 1920s to the 1950s: Class, Domesticity and Bohemianism* (Oxford: Oxford University Press, 2001), 11.
92. Ibid., 12.
93. Sayers, "Absolutely Elsewhere," 188.
94. Ibid., 192.
95. Ibid., 195. During Harcourt Grimbold's interrogation, Sayers gives the distance of the Lilacs from London as just over forty miles and the journey time as "an hour and a quarter from door to door," 190.

96. Ibid., 196.
97. Clock synchronization as a significant context for Einstein's work is discussed in Galison, *Einstein's Clocks*, chaps. 1 and 5.
98. The multiple facets of clock coordination are discussed in Galison, *Einstein's Clocks*, 40–41, 311–28.
99. Gordon Stokes, "The Spark and the Scythe: Modern Methods of Electric Time-Keeping," *Conquest* 2, no. 4 (February 1921): 176–80; F. Hope-Jones, "Greenwich Time by Wireless," *Conquest* 5, no. 5 (March 1924): 198–200; F. Addey, "How Time Signals Are Broadcast," *Discovery* 10, no. 111 (March 1929): 81–84; A. M. Low, "The Right Time—Please!," *Armchair Science* 2, no. 6 (September 1930): 324–27.
100. Plain, *Women's Fiction*, 47.
101. For an account of how canonical modernist authors navigated these same issues, see Stevenson, *Modernist Fiction*, chap. 3.
102. Beer, *Open Fields*, 187.
103. For a reevaluation of Sinclair's writings, see Andrew J. Kunka and Michele K. Troy, eds., *May Sinclair: Moving Towards the Modern* (Aldershot: Ashgate, 2006), and Leigh Wilson, "May Sinclair," in *The Literary Encyclopedia*, July 17, 2001, accessed October 14, 2009, http://www.litencyc.com/php/speople.php?rec=true&UID=4086.
104. May Sinclair, "The Finding of the Absolute," in *Uncanny Stories* (London: Hutchinson, 1923), 225–47. Sinclair's exchanges with T. S. Eliot are discussed by Rebecca Kinnamon Neff in " 'New Mysticism' in the Writings of May Sinclair and T. S. Eliot," *Twentieth Century Literature* 26, no. 1 (1980): 82–108.

CHAPTER SIX

1. MS Eng 1401, bMS 870 (folder 4), Empson Papers, Houghton Library, Harvard University.
2. MS Eng 1401, bMS 1032, Empson Papers, and John Haffenden, *Among the Mandarins*, vol. 1 of *William Empson* (Oxford: Oxford University Press, 2005), 126. On Cambridge undergraduate attitudes and activities during the General Strike, see T. E. B. Howarth, *Cambridge Between Two Wars* (London: Collins, 1978), 148–50.
3. MS Eng 1401, bMS 1032, Empson Papers. It is possible that Empson is using "hoisting stations" in a metaphorical sense to evoke the image of people clustering around to wait for news (for instance, at the head of a mine after an accident). Signal flags aboard ship are "hoists," and flags may have been run up in the market square to announce wireless broadcasts. I am grateful to Debby Banham, John Hart, and Simon Schaffer for suggested interpretations of this phrase.
4. H. G. Wells, *The Sleeper Awakes* (London: Collins, 1954; first published as *When the Sleeper Wakes*, 1899, revised 1910), 200.

5. Wells, *Sleeper Awakes*, 201.
6. MS Eng 1401, bMS 870 (folder 4), Empson Papers.
7. F. W. Payne, *John Donne and His Poetry* (London: Harrap: 1926), 164; Empson, *Essays on Renaissance Literature*, vol. 1, 129.
8. Empson, *Essays on Renaissance Literature*, vol. 1, 129.
9. I have discussed the ideas about space travel informing "Donne the Space Man" in relation to Empson's first published poem, "The Ants," in "William Empson, Ants and Aliens" in *Science and Modern Poetry: New Directions*, ed. John Robert Holmes (Liverpool: Liverpool University Press, 2012).
10. MS Eng 1401, bMS 870 (folder 4) Empson Papers.
11. MS Eng 1401, bMS 870 (folder 5) Empson Papers.
12. Empson, *Essays on Renaissance Literature*, vol. 1, 78–9.
13. MS Eng 1401, bMS 870 (folder 5), Empson Papers.
14. MS Eng 1401, bMS 870 (folder 4), Empson Papers. The "east wind" comment had been phrased less cuttingly in Empson, "The Style of the Master," in *T. S. Eliot: A Symposium*, ed. Richard March and Tambimuttu (London: Routledge, 1965; first published 1948), 35.
15. Empson, *Essays on Renaissance Literature*, vol. 1, 78.
16. This poem is discussed in detail in my essay "Flame far too hot."
17. Empson, *Complete Poems*, 205.
18. Ibid., 29.
19. Ibid., 205.
20. H. G. Wells, *The Time Machine*, ed. John Lawton, Everyman New Centennial edition (London: Dent, 1995; first published 1895), 17.
21. Empson, *Complete Poems*, 205.
22. Eddington, *Nature of the Physical World*, 50.
23. Eddington, *Stars and Atoms* (Oxford: Clarendon, 1927), 83.
24. Ibid., 51.
25. Ibid., 83.
26. Empson, *Essays on Renaissance Literature*, vol. 1, 84.
27. Empson, *Complete Poems*, 162.
28. Eddington, *Nature of the Physical World*, 82.
29. Ibid., 83.
30. Empson, *Complete Poems*, 13.
31. Ibid., 163.
32. Ibid., 13.
33. Eddington, *Nature of the Physical World*, 49.
34. Empson, *Complete Poems*, 164.
35. On *Experiment* magazine, see Jason Harding, "*Experiment* in Cambridge: 'A Manifesto of Young England,'" *Cambridge Quarterly* 27, no. 4 (1998): 287–309, and my online essay "Finite But Unbounded: *Experiment* Magazine, Cambridge, England, 1928–31," *Jacket* 20 (2002), accessed April 30, 2011, http://jacketmagazine.com/20/price-expe.html.
36. Harding, "*Experiment* in Cambridge," 289.

37. George Ireland, "Hare, William Francis, fifth earl of Listowel (1906–1997)," in *Oxford Dictionary of National Biography* (Oxford University Press, May 2010), accessed April 30, 2011, http://www.oxforddnb.com/view/article/65196.
38. Harding, "*Experiment* in Cambridge," 296.
39. "Songs for Sixpence," *Cambridge Gownsman and Undergraduette*, November 30, 1929, 22.
40. "Nothing Venture," by T. W., *Granta*, June 7, 1929, 531.
41. Varieties of aspidistra were brought to Britain from China and Japan during the nineteenth century, and the plant's ability to withstand gas fumes earned it the name "parlour palm." See Catherine Horwood, *Potted History: The Story of Plants in the Home* (London: Frances Lincoln, 2007), 105, 136, 159. By the early twentieth century, the aspidistra had become a ready symbol of aspirations to middle-class respectability, a connotation enshrined in George Orwell's *Keep the Aspidistra Flying* (1936).
42. *Gownsman*, January 26, 1929, 8.
43. *Gownsman*, January 26, 1929, 9.
44. "Home Thoughts from Abroad," *Gownsman*, January 17, 1931, 6. See note 35 for a Web link to the image.
45. Harding, "*Experiment* in Cambridge," 294.
46. "Congratulations to . . . ," *Gownsman*, March 1, 1930, 2; "The Cult of the Moustache," *Gownsman*, May 17, 1930, 5.
47. Haffenden, *Among the Mandarins*, chap. 9.
48. Empson, *Complete Poems*, 212; Haffenden, *Among the Mandarins*, 232–39, 351.
49. Haffenden, *Among the Mandarins*, 239.
50. I have discussed "Letter IV" in detail in "Monogamy and the Next Step?"
51. A sixth Letter poem looking back over the sequence was composed on the occasion of Lee's marriage in 1935 and published posthumously.
52. Empson, *Complete Poems*, 31.
53. Blaise Pascal, *Pensées*, III, 206 (Paris: G. Crès, 1924; first published 1670), 138. The complexities of this phrase and its significance to modern readers of Pascal are discussed by Elisabeth Marie Loevlie in *Literary Silences in Pascal, Rousseau, and Beckett* (Oxford: Clarendon Press, 2003), chap. 4.
54. Huxley, *Point Counter Point*, 129–30.
55. Ibid., 198.
56. Ibid., 224.
57. Sayers, *Documents in the Case*, 51.
58. Empson, *Complete Poems*, 31.
59. Empson, *Complete Poems*, 212; Haffenden, *Among the Mandarins*, 233.
60. J. B. S. Haldane, *Possible Worlds and Other Essays* (London: Chatto and Windus, 1927), 1.
61. Empson, *Complete Poems*, 31.
62. Ibid., 211–12. This edition has "ever" for "never," but collections published during Empson's lifetime show that he wrote "never"; see Empson,

Poems (London: Chatto & Windus, 1935), 45, and *Collected Poems* (London: Chatto & Windus, 1955), 100.

63. Debates on extraterrestrial life are placed in their Victorian context by Morus in *When Physics Became King*, 221–23. The transition from theological controversy over plurality of worlds to acceptance of "limited pluralism" is summarized in Gregory, "Intersections of Physical Sciences and Western Religion," 37–39.
64. "Einstein on the Mystic Wireless—What Martians Would Do," *Daily Mail*, January 31, 1920, 5.
65. Morus discusses proposals for light ray signaling between Earth and Mars in *When Physics Became King*, 221–22.
66. On *Aelita*, see Ian Christie, "Down to Earth: *Aelita* Relocated," in *Inside the Film Factory: New Approaches to Russian and Soviet Cinema*, eds. Richard Taylor and Ian Christie (London: Routledge, 1994), 80–102.
67. The signals received on August 24, 1924, are discussed by John A. Keel in *Our Haunted Planet* (London: Futura, 1975; first published 1971), 156–57.
68. Brisbane, "The End—and the Beginning."
69. J. B. S. Haldane, "The Last Judgment," in *Possible Worlds and Other Essays* (London: Chatto and Windus, 1927), 298.
70. Joseph H. Elgie, "The Map of Mars," *Conquest* 1, no. 6 (April 1920): 282.
71. H. Spencer Toy, "Life on Other Worlds," *Conquest* 4, no. 6 (April 1923): 224.
72. Eddington, *Nature of the Physical World*, 173–75.
73. Spencer Toy, "Life on Other Worlds," 227.
74. Ibid., 224, 225.
75. Low, "Can Science Explain Love?"
76. Macaulay, *Potterism*, 230.
77. Empson, *Complete Poems*, 31.
78. Quoted in Gordon Johnson, *University Politics: F. M. Cornford's Cambridge and His Advice to the Young Academic Politician* (Cambridge: Cambridge University Press, 1994), 63.
79. Empson, *Complete Poems*, 211.
80. Empson, *Some Versions of Pastoral* (London: Penguin, 1995; first published 1935), 67–68. Also quoted in Empson, *Complete Poems*, 215.
81. Empson, *Complete Poems*, 211.
82. Haffenden, in *Among the Mandarins*, chap. 13, discusses the central role played by this figure in Empson's subversive vision, formulated in *Some Versions of Pastoral* (1935), and its relationship to the scapegoat figure elaborated by James Frazer in *The Golden Bough* (1890).
83. Empson, *Complete Poems*, 205.
84. Eddington, *Stars and Atoms*, 83.
85. Various accounts of this confrontation are summarized in Stanley, *Practical Mystic*, 258.
86. See Simon Schaffer, "Where Experiments End: Tabletop Trials in Victorian

Astronomy," in *Scientific Practice: Theories and Stories of Doing Physics*, ed. Jed Z. Buchwald (Chicago: University of Chicago Press, 1995), 257–99.

87. An accessible account of these developments is given in Morus, *When Physics Became King*, 209–16. The significance of amateurs in early astrophysics is explored by John Lankford in "Amateurs and Astrophysics: A Neglected Aspect in the Development of a Scientific Specialty," *Social Studies of Science* 11, no. 3 (1981): 275–303.
88. Stanley, *Practical Mystic*, 53.
89. Ibid., 32, 44, 47–49.
90. Arthur Eddington, *The Internal Constitution of the Stars* (Cambridge: Cambridge University Press, 1926), 4.
91. Ibid., 6.
92. *Le Principe de Relativité et la Théorie de la Gravitation*, by Jean Bequerel, review by J. R., *Science Progress* 18, no. 70 (October 1923): 314.
93. K. S. Thorne, *Black Holes and Time Warps: Einstein's Outrageous Legacy* (New York: Norton, 1994), 134.
94. Eddington, *Internal Constitution of the Stars*, 6.
95. Eddington, *Stars and Atoms*, 83.
96. Opportunities for the "inappropriate reader" of science writing are discussed in Beer, *Open Fields*, 187.
97. Eddington, *Internal Constitution of the Stars*, 131.
98. Ibid., 171.
99. Ibid., 48. This subheading follows "The Story of Algol," which Eddington also refers to as the mystery of "The Missing Word and the False Clue."
100. Eddington, *Stars and Atoms*, 50.
101. Eddington, *Internal Constitution of the Stars*, 172.
102. Eddington, *Stars and Atoms*, 122.
103. Ibid., 125.
104. Ibid., 127.
105. These themes are explored in Beer, *Open Fields*, 219–41, and Clarke, *Energy Forms*. Victorian models for solar energy are discussed in Schaffer, "Where Experiments End," 283–99.
106. Empson, *Complete Poems*, 31.
107. Haldane, "The Last Judgment," 289.
108. Ibid., 290.
109. William Empson, "Almost," *Granta*, January 27, 1927, in *Empson in* Granta: *The Book, Film and Theatre Reviews of William Empson*, ed. Eric Griffiths (Tunbridge Wells: Foundling Press, 1993), 34.
110. Empson, "Almost," 35.
111. Empson, *Complete Poems*, 211.
112. Eddington, *Internal Constitution of the Stars*, 19–20.
113. John Haffenden describes Empson's upper second class degree in mathematics as a "disappointing result." Empson, *Complete Poems*, lxxi.

114. K. C. Wali, *Chandra: A Biography of S. Chandrasekhar* (Chicago: University of Chicago Press, 1991), chap. 5.
115. Ibid., 142.
116. Subramanyan Chandrasekhar, "The Highly Collapsed Configurations of a Stellar Mass," *Monthly Notices of the Royal Astronomical Society* 91 (1931): 466.
117. MS Eng 1401, bMS 1401/1145 (folder 5), Empson Papers.
118. K. S. Thorne, "The Search for Black Holes," *Scientific American* 231, no. 6 (December 1974): 32.
119. MS Eng 1401, bMS 870 (folder 4), Empson Papers.
120. For an example of such insight on the part of a scientist making use of creativity and social awareness, see the discussion of Frederick Soddy's warning about atomic war in 1915 in Sclove, "From Alchemy to Atomic War."

CONCLUSION

1. O'Connor, "Reflections on Popular Science in Britain," 338.
2. Ibid., 338–39.
3. For studies of popular reception in Germany and France, see Lewis Elton, "Einstein, General Relativity and the German Press, 1919–1920," *Isis* 77 (1986): 95–103, and Michel Biezunski, "Popularization and Scientific Controversy: The Case of the Theory of Relativity in France," in *Expository Science: Forms and Functions of Popularisation*, eds. Terry Shinn and Richard Whitley (Dordrecht: Reidel, 1985). Michel Biezunski, in "Einstein's Reception in Paris in 1922" in Thomas F. Glick, *The Comparative Reception of Relativity* (Dordrecht: Reidel, 1987), 169–88, also includes a brief discussion of popular responses to Einstein in France; in the same volume, Barbara J. Reeves, in "Einstein Politicized: The Early Reception of Relativity in Italy," 189–229, explores the public politicization of relativity in Italy in some depth; Tsutomu Kaneko, in "Einstein's Impact on Japanese Intellectuals," 351–79, focuses on Japanese intellectual resistance to Einstein; and Glick, in "Relativity in Spain," 231–63, provides an overview of scientific and popular reception in Spain, enlarged in *Einstein in Spain: Relativity and the Recovery of Science* (Princeton, NJ: Princeton University Press, 1988). B. Bensaude-Vincent explores the philosophical reception of relativity in "When a Physicist Turns on Philosophy: Paul Langevin (1911–1932)," *Journal of the History of Ideas* 49 (1988): 319–38. Richard A. Jarrell focuses on scientific responses in Canada, with a brief mention of popular reception, in "The Reception of Einstein's Theory of Relativity in Canada," *Journal of the Royal Society of Astronomy of Canada* 73 (1979): 358–69. Danian Hu, in *China and Albert Einstein: The Reception of the Physicist and His Theory in China 1917–1979* (Cambridge, MA: Harvard University Press, 2005), is largely focused on scientific reception in China. María de la Paz Ramos Lara discusses reception among scientists in "The Reception of Relativity

in Mexico," *Synthesis Philosophica* 42 (2006): 299–304. Adel Ziadat discusses popularization of relativity in the Arab world in "Early Reception of Einstein's Relativity in the Arab Periodical Press," *Annals of Science* 51 (1994): 17–35.

4. Clark, *Einstein*, 267.
5. A helpful introduction to Dunne and his literary reception is given by Victoria Stewart, "J. W. Dunne and Literary Culture in the 1930s and 1940s," *Literature and History* 17 (2008): 62–81.
6. Lewis Pyenson, Sean Johnston, Alberto Martínez, and Richard Staley, "Revisiting the History of Relativity," *Metascience* 20 (2011): 53–73.

Bibliography

Unsigned Newspaper and Magazine Articles

"Arithmetic Extraordinary." *Daily Mail,* November 8, 1919, 6.

"Astronomers on Einstein—New Geometry Wanted—Dr. Eddington and Relativity." *Times,* December 13, 1919, 9.

"Bent Light Puzzle—Euclid Not 'Caught Out.'" *Daily Mail,* November 8, 1919, 5.

"Books of the Month." *Discovery* 1, no. 9 (September 1920): 271–72.

"Breaking Up the Atom: Professor Rutherford's Discoveries." *Daily Mail,* June 11, 1921, 5.

"Charivaria." *Punch,* June 8, 1921, 441.

"Charivaria." *Punch,* June 22, 1921, 481.

"Cinemativity—Time and Space as Puppets of the Screen." *Daily Express,* June 14, 1921, 5.

"Congratulations to . . ." *Gownsman,* March 1, 1930, 2.

"Crowds to Hear Einstein—'Relativity' Vies With Cricket." *Daily News,* June 14, 1921, 1.

"The Cult of the Moustache." *Gownsman,* May 17, 1930, 5.

"The Eclipse." *London Magazine* 44, no. 113 (March 1920): 65–68.

"Einstein." *Times,* December 4, 1919, 15.

"Einsteinized." *Punch,* December 10, 1919, 498.

"Einstein on His Theory—Relativity and Philosophy." *Times,* June 13, 1921, 11.

"Einstein on His Theory: Time, Space and Gravitation, The Newtonian System." *Times,* November 28, 1919, 13–14.

"Einstein on the Mystic Wireless—What Martians Would Do." *Daily Mail,* January 31, 1920, 5.

"Einstein's Debut." *Punch,* June 22, 1921, 490.

"The Einstein Upheaval." *Punch,* November 26, 1919, 442.

"The Ether of Space—Sir Oliver Lodge's Caution." *Times*, November 8, 1919, 12.

"Euclid Up-to-Date—Novel Possibilities in the 'Fourth Dimension.'" *Daily Sketch*, November 10, 1919, 2.

"The Fabric of the Universe." *Times*, November 7, 1919, 13.

"'A Fashion of To-Day.'" *Times*, December 2, 1919, 17.

"Food Prices Still Higher." *Times*, November 7, 1919, 13.

Four Lectures on Relativity and Space, by C. P. Steinmetz. Review by J. W. F. *Science Progress* 18, no. 72 (1924): 651–52.

"The Fourth Dimension." *Morning Post*, November 11, 1919, 8.

"High Milk Prices." *Times*, November 7, 1919, 13.

"How Tall Are You, Einstein Measure? Prof. Eddington, 6 Feet to the Eye, Explains How It May Be Really Only 3 Feet." *New York Times*, December 4, 1919, 19.

"In Our Opinion." *Armchair Science* 1, no. 11 (February 1930): 646–47.

"'It All Depends.'" *Times*, May 27, 1921, 11.

"Jazz in Scientific World." *New York Times*, November 16, 1919, sec. 3, 8.

"Labour Disputes Courts—Second Reading of the Bill." *Times*, November 7, 1919, 12.

"The Last Strawinksy." *Daily News*, June 11, 1921, 5.

"Leeds in a Bad Way." *Daily Mail*, November 8, 1919, 5.

Letters to the Editor. *Daily Mail*, November 8, 1919, 5.

"Light and Logic." *New York Times*, November 16, 1919, sec. 3, 1.

"Light Caught Bending—A Discovery Like Newton's." *Daily Mail*, November 7, 1919, 7.

"Light on the Bend." *Morning Post*, November 8, 1919, 7.

"Lights All Askew in the Heavens." *New York Times*, November 10, 1919, 17.

"Light Weighed—Greatest Discovery Since That of Gravitation." *Daily Express*, November 7, 1919, 7.

The Mathematical Theory of Relativity, by A. Kopff. Review by B. S. *Science Progress* 18, no. 72 (1924): 647.

"Motor Show—Opening To-Day—More Cars for Women." *Times*, November 7, 1919, 13.

"Music and Science." *Daily News*, June 15, 1921, 4.

"Nothing Venture." by W. T. *Granta*, June 7, 1929, 531.

"The Past and the Future." *Times*, November 22, 1919, 13.

Le Principe de Relativité et la Théorie de la Gravitation, by Jean Bequerel. Review by J. R. *Science Progress* 18, no. 70 (October 1923): 313–14.

"The Principle of Relativity." *Science Progress* 9, no. 34 (October 1914): 352–53.

"Relativity." *Armchair Science* 1, no. 11 (February 1930): 652–54.

"Relativity Chase in a Ship—How Professor Einstein Arrived." *Daily Express*, June 9, 1921, 5.

Report of Meeting November 1918. *Journal of the British Astronomical Association* 29, no. 2 (November 1918): 35–39.

"The Revolution in Science—Astronomers' Discussion." *Times*, November 15, 1919, 13–14.

"Revolution in Science—New Theory of the Universe." *Times*, November 7, 1919, 12.
"The Revolution in Science—Reconstruction Proposed by Sir J. Larmor." *Times*, November 21, 1919, 12.
"The Run Home: An Einsteinist at Large." *Times*, December 2, 1919, 17.
"Silence and Remembrance." *Times*, November 7, 1919, 13.
"Sir O. Lodge on Einstein's Theory." *Times*, November 25, 1919, 16.
"Some Einstein Perplexities—Danger Posts in Path of Little Knowledge." *Daily News*, June 11, 1921, 5.
"Songs for Sixpence." *Cambridge Gownsman and Undergraduette*, November 30, 1929, 22.
The Theory of Relativity, by Erwin Freundlich. Review. *Science Progress* 18, no. 72 (1924): 648.
"Unpopular Science." *Daily News*, November 8, 1919, 6.
"Upsetting the Universe—Dizzy Results of the New Light Discovery." *Daily Express*, November 8, 1919, 5.
"Wanted: A Popular Science Picture Theatre." *Conquest* 3, no. 10 (August 1922): 423.
"'Which Is Absurd,'" *Daily Express*, November 8, 1919, 4.
"'Which Is Absurd,'" *Daily Express*, November 10, 1919, 6.
"Why Nothing Stands Still." *Tit-Bits*, March 5, 1921, 7.

Books and Articles

Albright, Daniel. *Quantum Poetics: Yeats, Pound, Eliot and the Science of Modernism*. Cambridge: Cambridge University Press, 1997.
Ashford, A. E. "The Q-Ray." *Red Magazine* 44, no. 263 (July 9, 1920): 195–203.
———. "The Time-Adjuster." *Red Magazine* 40, no. 334 (March 30, 1923): 325–32.
Ashley, Mike. *The Age of the Storytellers: British Popular Fiction Magazines 1880–1950*. London: British Library, 2006.
———. *Time Machines: The Story of the Science Fiction Pulp Magazines from the Beginning to 1950*. Liverpool: Liverpool University Press, 2000.
Auden, W. H., and Louis MacNeice. *Letters from Iceland*. London: Faber, 1937.
Avery, Todd. *Radio Modernism: Literature, Politics and the BBC, 1922–1938*. Aldershot: Ashgate, 2006.
Bagwell, Philip S. *The Railwaymen: The History of the National Union of Railwaymen*. London: George Allen & Unwin, 1963.
Bailey Ogilvie, Marilyn. "Obligatory Amateurs: Annie Maunder (1868–1947) and British Women Astronomers at the Dawn of Professional Astronomy." *British Journal for the History of Science* 33, no. 1 (2000): 67–84.
Banfield, Ann. *The Phantom Table: Woolf, Fry, Russell, and the Epistemology of Modernism*. Cambridge: Cambridge University Press, 2000.

[Barton Kelly, T.] "Notice to Readers." *Conquest* 7, no. 5 (March 1926) 126.

Beer, Gillian. *Darwin's Plots: Evolutionary Narrative in Darwin, George Eliot and Nineteenth-Century Fiction*. London: Routledge & Kegan Paul, 1983.

———. "Eddington and the Idiom of Modernism." In *Science, Reason, and Rhetoric*, edited by Henry Krips, J. E. McGuire, and Trevor Melia, 295–315. Pittsburgh: University of Pittsburgh Press, 1995.

———. *Open Fields: Science in Cultural Encounter*. Oxford: Oxford University Press, 1996.

———. *Wave, Atom, Dinosaur: Woolf's Science*. London: Virginia Woolf Society, 2000.

———. "'Wireless': Popular Physics, Radio and Modernism." In *Cultural Babbage: Technology, Time, and Invention*, edited by Francis Spufford and Jenny Uglow, 149–66. London: Faber, 1996.

Bell, G. K. A. *Randall Davidson, Archbishop of Canterbury*. Oxford: Oxford University Press, 1935.

[Benn, John A.] "Editorial Notes." *Discovery* 10, no. 111 (March 1929): 71–72.

Bennett, Arnold. "The Angry Mr. Mencken: 'Lacks Balance, Sense of Justice and Evidence—But Has Done Good'" (March 31, 1927). In *The Evening Standard Years: "Books and Persons" 1926–1931*, edited by Andrew Mylett, 36–38. London: Chatto and Windus, 1974.

———. "Einstein for the Tired Business Man" (April 21, 1927). In *The Evening Standard Years: "Books and Persons" 1926–1931*, edited by Andrew Mylett, 42–44. London: Chatto and Windus, 1974.

———. *The Evening Standard Years: "Books and Persons" 1926–1931*. Edited by Andrew Mylett. London: Chatto and Windus, 1974.

———. "The 'Monstrous Conceit' of Some Modernists" (1 March 1928). In *The Evening Standard Years: "Books and Persons" 1926–1931*, edited by Andrew Mylett, 131–33. London: Chatto and Windus, 1974.

———. *The Night Visitor, and Other Stories*. London: Cassell, 1931.

Bensaude-Vincent, B. "When a Physicist Turns on Philosophy: Paul Langevin (1911–1932)." *Journal of the History of Ideas* 49 (1988): 319–38.

Beresford, Leslie. "The Stranger from Somewhere." *Red Magazine* 41, no. 345 (August 31, 1923): 476–85.

Biezunski, Michel. "Einstein's Reception in Paris in 1922." In *The Comparative Reception of Relativity*, edited by Thomas F. Glick, 169–88. Dordrecht: Reidel, 1987.

———. "Popularization and Scientific Controversy: The Case of the Theory of Relativity in France." In *Expository Science: Forms and Functions of Popularisation*, edited by Terry Shinn and Richard Whitley. Sociology of the Sciences 9. Dordrecht: Reidel, 1985.

Biggs, H. F. "The Electron and Professor Eddington." *Discovery* 10, no. 112 (April 1929): 132–34.

———. "What the New Relativity Means." *Discovery* 10, no. 114 (June 1929): 196–200.

Boccardi, M. G. "German Science and Latin Science." *Observatory* no. 504 (September 1916): 381–91.

Boon, Tim. *Films of Fact: A History of Science in Documentary Films and Television*. London: Wallflower, 2008.

Bowler, Peter. *Charles Darwin: The Man and His Influence*. Cambridge, MA: Blackwell, 1990.

———. "Experts and Publishers: Writing Popular Science in Early Twentieth-Century Britain, Writing Popular History of Science Now." *British Journal for the History of Science* 39, no. 2 (2006): 159–87.

———. *Reconciling Science and Religion: The Debate in Early-Twentieth-Century Britain*. Chicago: University of Chicago Press, 2001.

———. *Science For All: The Popularization of Science in Early Twentieth-Century Britain*. Chicago: University of Chicago Press, 2009.

Bradshaw, David. "The Best of Companions: J. W. N. Sullivan, Aldous Huxley, and the New Physics. [Part One]." *Review of English Studies*, n.s. 47, no. 186 (1996): 188–206.

———. "The Best of Companions: J. W. N. Sullivan, Aldous Huxley and the New Physics. [Part Two]." *Review of English Studies*, n.s. 47, no. 187 (1996): 352–68.

Brazier, L. G., "What Is the Use of Einstein?" *Conquest* 3, no. 3 (January 1922): 123–24.

Brians, Paul. *Nuclear Holocausts: Atomic War in Fiction*. Kent: Kent State University Press, 1987.

Briggs, Asa. *The History of Broadcasting in the United Kingdom*. Vol. 1, *The Birth of Broadcasting*. Oxford: Oxford University Press, 1995.

Brisbane, Coutts. "The Almighty Atom." *Red Magazine* 50, no. 307 (March 17, 1922): 611–21.

———. "An Elementary Affair." *Red Magazine* 57, no. 350 (November 9, 1923): 259–70.

———. "The End—and the Beginning." Part 1, *Yellow Magazine* 15, no. 99 (June 26, 1925): 689–701; part 2, *Yellow Magazine* 16, no. 100 (July 10, 1925): 37–48.

———. "Mr. Fipkins and the Younger World." *Red Magazine* 57, no. 348 (October 12, 1923): 77–85.

———. "Thus Said Pel!" *Red Magazine* 44, no. 263 (July 9, 1920): 209–24.

Broks, Peter. *Media Science before the Great War*. Basingstoke: Macmillan, 1996.

Brown, Daniel. *Hopkins' Idealism: Philosophy, Physics, Poetry*. Oxford: Clarendon Press, 1997.

Buchan, John. *The Gap in the Curtain*. London: Hodder and Stoughton, 1932.

Burrows, Arthur R. "Juggling With Air." *Conquest* 1, no. 3 (January 1920): 103–10.

Butts, Mary. *The Journals of Mary Butts*. Edited by Nathalie Blondel. New Haven: Yale University Press, 2002.

Cain, Sarah. "The Metaphorical Field: Post-Newtonian Physics and Modernist Literature." *Cambridge Quarterly* 28 (1999): 46–64.

Carroll, Lewis. *The Annotated Alice*. Edited by Martin Gardner. London: Penguin, 2001.

Case, Thomas. *Letters to the* Times, *1884–1922*. Edited by. R. B. Mowat. Oxford: John Johnson, 1927.

———. "Theories of Space—Newton and Einstein—The Absolute and the Relative." Letter to the editor, *Times*, November 22, 1919, 8.

Catterall, Peter, Colin Seymour-Ure, and Adrian Smith, eds. *Northcliffe's Legacy: Aspects of the British Popular Press, 1896–1996*. Basingstoke: Macmillan, 2000.

Chandrasekhar, Subramanyan. "The Highly Collapsed Configurations of a Stellar Mass." *Monthly Notices of the Royal Astronomical Society* 91 (1931): 456–66.

Chisholm, Anne, and Michael Davie. *Beaverbrook: A Life*. London: Hutchinson, 1992.

Christie, Ian. "Down to Earth: *Aelita* Relocated." In *Inside the Film Factory: New Approaches to Russian and Soviet Cinema*, edited by Richard Taylor and Ian Christie, 80–102. London: Routledge, 1994.

Clark, Ronald. *Einstein: The Life and Times*. London: Hodder and Stoughton, 1979.

Clarke, Bruce. *Energy Forms: Allegory and Science in the Era of Classical Thermodynamics*. Ann Arbor: University of Michigan Press, 2001.

Clay, Catherine. *British Women Writers 1914–1945: Professional Work and Friendship*. Aldershot: Ashgate, 2006.

Collier, Patrick. *Modernism on Fleet Street*. Aldershot: Ashgate, 2006.

Collins, Paul, ed. *MUP Encyclopaedia of Australian Science Fiction and Fantasy*. Carlton, Vic.: Melbourne University Press, 1998.

Cooter, Roger, and Stephen Pumfrey. "Separate Spheres and Public Places: Reflections on the History of Science Popularization and Science in Popular Culture." *History of Science* 32 (1994): 237–67.

Craige, Betty Jean. *Literary Relativity: An Essay on Twentieth-Century Narrative*. Lewisburg: Bucknell University Press, 1982.

Crawford, Alice. *Paradise Pursued: The Novels of Rose Macaulay*. Madison, NJ: Fairleigh Dickinson University Press, 1995.

Crommelin, A. C. D. "Astronomy Widens Its Vision." *Discovery* 10, no. 111 (March 1929): 78–80.

Davidson, Charles. "Weighing Light." *Conquest* 1, no. 3 (January 1920): 123–32.

Dearden, R. L. "The Morning Star." *Novel Magazine* 40 (January 1925): 315–18.

de la Paz Ramos Lara, María. "The Reception of Relativity in Mexico." *Synthesis Philosophica* 42 (2006): 299–304.

de Tunzelman, G. W. "The General Theory of Relativity and Einstein's Theory of Gravitation." *Science Progress* 13, no. 52 (April 1919): 652–57.

———. "Physical Relativity Hypotheses Old and New." *Science Progress* 13, no. 51 (January 1919): 475–82.

DiBattista, Maria, and Lucy McDiarmid, eds. *High and Low Moderns: Literature and Culture, 1889–1939*. Oxford: Oxford University Press, 1996.

Donaldson, Lewis. "Deans as Class Warriors." *New Leader*, March 6, 1925, 3–4.

Dorril, Stephen. *Blackshirt: Sir Oswald Mosley and British Fascism*. London: Viking, 2006.

Douglas, C. H. "Credit-Power and Democracy—Chapter 12." *New Age* 27, no. 13 (July 29, 1920): 196.

Dunne, J. W. "Can We Travel in the Time Dimension?" *Armchair Science* 3, no. 7 (October 1931): 415–16.

———. *An Experiment with Time*. London: A and C Black, 1927.

Earle, David M. *Re-covering Modernism: Pulps, Paperbacks, and the Prejudice of Form*. Farnham: Ashgate, 2009.

Eddington, Arthur Stanley. "The Domain of Physical Science." In *Science, Religion, and Reality*, ed. Joseph Needham, 187–218. London: Sheldon, 1925.

———. "Einstein on Time and Space." *Quarterly Review* no. 462 (January 1920): 226–36.

———. "Einstein's Theory of Space and Time." *Contemporary Review* 116, no. 648 (December 1919): 639–43.

———. "The Future of International Science." *Observatory* no. 501 (June 1916): 270–72.

———. "Gravitation and the Principle of Relativity." *Nature* 101, no. 2523 (March 7, 1918): 15–17; no. 2524 (March 14, 1918): 34–36.

———. *The Internal Constitution of the Stars*. Cambridge: Cambridge University Press, 1926.

———. "The Meaning of Matter and the Laws of Nature According to the Theory of Relativity." *Mind* 29 (1920): 145–58.

———. *The Nature of the Physical World*. Cambridge: Cambridge University Press, 1928.

———. "The Philosophical Aspect of the Theory of Relativity." *Mind* 29 (1920): 415–22.

———. *The Philosophy of Physical Science*. Cambridge: Cambridge University Press, 1939.

———. *Space Time and Gravitation*. Cambridge: Cambridge University Press, 1920.

———. *Stars and Atoms*. Oxford: Clarendon, 1927.

———. *The Theory of Relativity and Its Influence on Scientific Thought*. Oxford: Clarendon, 1922.

Edgell, Derek. *The Order of Woodcraft Chivalry, 1916–1949, as a New Age Alternative to the Boy Scouts*. Lewiston: Mellen, 1992.

Eger, Elizabeth, and Lucy Peltz. *Brilliant Women: 18th-Century Bluestockings*. London: National Portrait Gallery, 2008.

Einstein, Albert. "Einstein on His Theory: Time, Space and Gravitation, The Newtonian System." *Times*, November 28, 1919, 13–14.

———. *Relativity: The Special and the General Theory*. Trans. Robert W. Lawson. London: Routledge, 1960. First published 1920 by Methuen.

Elgie, Joseph H. "The Map of Mars." *Conquest* 1, no. 6 (April 1920): 282–84.

Eliot, T. S. *Complete Poems and Plays*. London: Faber, 1969.

———. "Religion and Science: A Phantom Dilemma." *Listener*, March 23, 1932, 428–29.

Elton, Lewis. "Einstein, General Relativity and the German Press, 1919–1920." *Isis* 77 (1986): 95–103.

Emmer, Michele, ed. *The Visual Mind II*. Cambridge, MA: MIT Press, 2005.

Empson, William. "Almost." *Granta*, January 27, 1927. In *Empson in* Granta: *The Book, Film and Theatre Reviews of William Empson*, ed. Eric Griffiths. Tunbridge Wells: Foundling Press, 1993.

———. *Collected Poems*. London: Chatto & Windus, 1955.

———. *Complete Poems*. Edited by John Haffenden. London: Allen Lane, 2000.

———. *Essays on Renaissance Literature*. Vol. 1, *Donne and the New Philosophy*. Edited by John Haffenden. Cambridge: Cambridge University Press, 1993.

———. *Poems*. London: Chatto & Windus, 1935.

———. *Some Versions of Pastoral*. London: Penguin, 1995. First published 1935.

———. "The Style of the Master." In *T. S. Eliot: A Symposium*, edited by Richard March and Tambimuttu, 35–37. London: Routledge, 1965. First published 1948.

Engels, Eve-Marie, and Thomas Glick, eds. *The Reception of Charles Darwin in Europe*. London: Continuum, 2008.

Ertz, Susan. "Relativity and Major Rooke." *Harper's Magazine* 148 (April 1924): 669–76.

Fleming, J. A. "Some Difficulties in the Theory of Relativity." *Conquest* 3, no. 10 (August 1922): 419–22.

Foster Hartman, Lee. "The Altar of Destiny." *Novel Magazine* 37 (August 1923): 408–15.

Frank, Philipp. *Einstein: His Life and Times*. Translated by George Rosen. New York: Knopf, 1947.

Friedman, Alan J. and Carol C. Donley. *Einstein as Myth and Muse*. Cambridge: Cambridge University Press, 1985.

Fussell, Paul. *The Great War and Modern Memory*. Oxford: Oxford University Press, 1975.

Galison, Peter. *Einstein's Clocks and Poincaré's Maps: Empires of Time*. London: Sceptre, 2003.

Geldof, Koenraad. "Look Who's Talking: Caliban in Shakespeare, Renan and Guéhenno." In *Constellation Caliban: Figurations of a Character*, edited by Nadia Lie and Theo d'Haen, 81–112. Amsterdam: Rodopi, 1997.

Glick, Thomas F., ed. *The Comparative Reception of Darwinism*. Austin: University of Texas Press, 1974.

———, ed. *The Comparative Reception of Relativity*. Boston Studies in the Philosophy of Science, 103. Dordrecht: Reidel, 1987.

———. "Relativity in Spain." In *The Comparative Reception of Relativity*, edited by Thomas F. Glick, 231–63. Dordrecht: Reidel, 1987.

———. *Einstein in Spain: Relativity and the Recovery of Science*. Princeton, NJ: Princeton University Press, 1988.

Gold, Barri. *Thermopoetics: Energy in Victorian Literature and Science*. Cambridge, MA: MIT Press, 2010.

Gooday, Graeme. *Domesticating Electricity: Technology, Uncertainty and Gender, 1880–1914*. London: Pickering and Chatto, 2008.

———. "Illuminating the Expert-Consumer Relationship in Domestic Electricity." In *Science in the Marketplace: Nineteenth-Century Sites and Experiences*, eds. Aileen Fyfe and Bernard Lightman, 231–68. Chicago: University of Chicago Press, 2007.

Gossin, Pamela. *Thomas Hardy's Novel Universe: Astronomy, Cosmology and Gender in the Post-Darwinian World*. Aldershot: Ashgate, 2007.

Graham, Loren. *Between Science and Values*. New York: Columbia University Press, 1981.

Graham Hall, Jean, and Douglas F. Martin. *Haldane: Statesman, Lawyer, Philosopher*. Chichester: Barry Rose Law, 1996.

Gregory, Frederick. "Intersections of Physical Sciences and Western Religion in the Nineteenth and Twentieth Centuries." In *Cambridge History of Science*, vol. 5, *The Modern Physical and Mathematical Sciences*, ed. Mary Jo Nye, 36–53. Cambridge: Cambridge University Press, 2003.

Griffiths, Dennis, ed. *Encyclopedia of the British Press, 1422–1992*. Basingstoke: Macmillan, 1992.

Haffenden, John. Introduction to *Complete Poems*, by William Empson, edited by John Haffenden. London: Allen Lane, 2000.

———. Introduction to *Essays on Renaissance Literature*, vol. 1, *Donne and the New Philosophy*, by William Empson, edited by John Haffenden. Cambridge: Cambridge University Press, 1993.

———. *William Empson*. Vol. 1, *Among the Mandarins*. Oxford: Oxford University Press, 2005.

———. *William Empson*. Vol. 2, *Against the Christians*. Oxford: Oxford University Press, 2006.

Haldane, J. B. S. "The Last Judgment." In *Possible Worlds and Other Essays*, 287–312. London: Chatto and Windus, 1927.

———. *Possible Worlds and Other Essays*. London: Chatto and Windus, 1927.

Haldane, J. R. "Synthetic Relativity." *Science Progress* 18, no. 69 (July 1923): 133–35.

Haldane, Richard Burdon, Viscount. *The Reign of Relativity*. London: John Murray, 1921.

Halsey, A. H. "Wootton, Barbara Frances, Baroness Wootton of Abinger (1897–1988)." In *Oxford Dictionary of National Biography*. Oxford University Press, October 2009. Accessed October 21, 2009. http://www.oxforddnb.com/view/article/39876.

Harding, Jason. "*Experiment* in Cambridge: 'A Manifesto of Young England.'" *Cambridge Quarterly* 27 no. 4 (1998): 287–309.

[Harris, Percy] "The Editor's Chair." *Conquest* 1, no. 2 (December 1919): 71.

———. "The Editor's Chair." *Conquest* 1, no. 3 (January 1920): 134.

———. "The Editor's Chair—A Short Talk on the Aims and Ideals of 'Conquest,'" *Conquest* 1, no. 1 (November 1919): 13.

———. "Questions about Einstein." *Conquest* 3, no. 7 (May 1922): 301.

———. "Questions and Answers." *Conquest* 1, no. 5 (March 1920): 247.

———. "Questions and Answers." *Conquest* 3, no. 4 (February 1922): 169.

———. "Questions and Answers—Solutions of Readers' Difficulties." *Conquest* 2, no. 5 (March 1921): 233–34.

———. "Some Strange Questions and a Few Experiments." *Conquest* 2, no. 10 (August 1921): 421–22.

Harrison, Frederic. "The Theory of Space—Practical Certainty and Relative Truth." Letter to the editor, *Times*, November 21, 1919, 8.

Harrow, Benjamin. *From Newton to Einstein: Changing Conceptions of the Universe.* London: Constable, 1920.

Henchman, Anna. "'The Globe we groan in': Astronomical Distance and Stellar Decay in *In Memoriam.*" *Victorian Poetry* 41 (2003): 29–45.

———. "Hardy's Stargazers and the Astronomy of Other Minds." *Victorian Studies* 51 (2008): 37–64.

Henderson, Linda Dalrymple. "Einstein and 20th-Century Art: A Romance of Many Dimensions." In *Einstein for the 21st Century: His Legacy in Science, Art, and Modern Culture*, edited by Peter L. Galison, Gerald Holton, and Silvan S. Schweber, 101–29. Princeton: Princeton University Press, 2008.

———. "Four-Dimensional Space or Space-Time? The Emergence of the Cubism-Relativity Myth in New York in the 1940s." In *The Visual Mind II*, edited by Michele Emmer, 349–97. Cambridge, MA: MIT Press, 2005.

———. *The Fourth Dimension and Non-Euclidean Geometry in Modern Art.* Second edition with a revised introduction. Cambridge, MA: MIT Press, forthcoming. Originally published 1983.

Henry, Holly. *Virginia Woolf and the Discourse of Science: The Aesthetics of Astronomy.* Cambridge: Cambridge University Press, 2003.

Herbert, Christopher. *Victorian Relativity: Radical Thought and Scientific Discovery.* Chicago: University of Chicago Press, 2001.

Hodgson, Peter E. "Relativity and Religion: The Abuse of Einstein's Theory." *Zygon* 38 no. 2 (2003): 393–409.

Horner, Frances. *Time Remembered.* London: Heinemann, 1933.

Horwood, Catherine. *Potted History: The Story of Plants in the Home.* London: Frances Lincoln, 2007.

Hostettler, John, and Brian P. Block. *Voting in Britain: A History of the Parliamentary Franchise.* Chichester: Barry Rose, 2001.

Howarth, T. E. B. *Cambridge Between Two Wars.* London: Collins, 1978.

Hu, Danian. *China and Albert Einstein: The Reception of the Physicist and His Theory in China 1917–1979.* Cambridge, MA: Harvard University Press, 2005.

Hughes, Jeff. "Radioactivity and Nuclear Physics." In *Cambridge History of Sci-*

ence, vol. 5, *The Modern Physical and Mathematical Sciences*, ed. Mary Jo Nye, 350–74. Cambridge: Cambridge University Press, 2003.

Humble, Nicola. *The Feminine Middlebrow Novel, 1920s to the 1950s: Class, Domesticity and Bohemianism*. Oxford: Oxford University Press, 2001.

Huxley, Aldous. *Crome Yellow*. London: Granada, 1982. First published 1921.

———. *Point Counter Point*. Harmondsworth: Penguin, 1975. First published 1928.

Ireland, George. "Hare, William Francis, fifth earl of Listowel (1906–1997)." *Oxford Dictionary of National Biography*. Oxford University Press, May 2010. Accessed April 30, 2011. http://www.oxforddnb.com/view/article/65196.

James, William. *Varieties of Religious Experience: A Study in Human Nature*. London: Routledge, 2002, centenary edition. First published 1902.

Jammer, Max. *Einstein and Religion: Physics and Theology*. Princeton: Princeton University Press, 1999.

Jarrell, Richard A. "The Reception of Einstein's Theory of Relativity in Canada." *Journal of the Royal Society of Astronomy of Canada* 73 (1979): 358–69.

Jeans, James. *The Mysterious Universe*. Harmsworth: Penguin, 1938.

Jenkins, Roy. *Asquith*. 3rd ed. London: Collins, 1986. First published 1964.

Johnson, Gordon. *University Politics: F. M. Cornford's Cambridge and His Advice to the Young Academic Politician*. Cambridge: Cambridge University Press, 1994.

Kaneko, Tsutomu. "Einstein's Impact on Japanese Intellectuals." In *The Comparative Reception of Relativity*, edited by Thomas F. Glick, 351–79. Dordrecht: Reidel, 1987.

Keel, John A. *Our Haunted Planet*. London: Futura, 1975. First published 1971.

Kenner, Hugh. *A Sinking Island: The Modern English Writers*. London: Barrie & Jenkins, 1988.

Kenney, Catherine. *The Remarkable Case of Dorothy L. Sayers*. Kent, OH: Kent State University Press, 1990.

Kent, Candice. " 'How does the mind move to Einstein's physics?': Science in the Writings of Virginia Woolf and Mary Butts," in *Restoring the Mystery of the Rainbow: Literature's Refraction of Science*, eds. Valeria Tinkler-Villani and C. C. Barfoot, 567–83. Amsterdam: Rodopi, 2011.

Kern, Stephen. *The Culture of Time and Space: 1880–1918*. Cambridge, MA: Harvard University Press, 1983.

Kibble, Matthew. " 'The Betrayers of Language': Modernism and the *Daily Mail*." *Literature and History*, 3rd ser., 11, no. 1 (2002): 62–80.

Kilmister, Clive. *Eddington's Search for a Fundamental Theory: A Key to the Universe*. Cambridge: Cambridge University Press, 2005.

Kinnamon Neff, Rebecca. " 'New Mysticism' in the Writings of May Sinclair and T. S. Eliot." *Twentieth Century Literature* 26, no. 1 (1980): 82–108.

Koss, Stephen. *Fleet Street Radical: A. G. Gardiner and the* Daily News. London: Allen Lane, 1973.

Kunka, Andrew J., and Michele K. Troy, eds. *May Sinclair: Moving Towards the Modern*. Aldershot: Ashgate, 2006.

Langevin, Paul. "L'evolution de l'espace et du temps." *Scientia* 10 (1911): 31–54.

Lankford, John. "Amateurs and Astrophysics: A Neglected Aspect in the Development of a Scientific Specialty." *Social Studies of Science* 11, no. 3 (1981): 275–303.

Larmor, Joseph. "The Einstein Theory—A Belgian Professor's Investigations." Letter to the editor, *Times*, January 7, 1920, 8.

Latour, Bruno. "A Relativistic Account of Einstein's Relativity." *Social Studies of Science* 18 (1988): 3–44.

Laugharne-Thornton, D. "The Einstein Theory of Relativity." *Science Progress* 16, no. 64 (April 1922): 641–43.

Lawrence, Christopher, and Anna-K Mayer, eds. *Regenerating England: Science, Medicine and Culture in Inter-war Britain*. Amsterdam: Rodopi, 2000.

Leane, Elizabeth. *Reading Popular Physics: Disciplinary Skirmishes and Textual Strategies*. Aldershot: Ashgate, 2007.

Leggett, H. W. "Whispers on the Wind." *Novel Magazine* 37 (August 1923): 416–21.

Leonardi, Susan J. *Dangerous by Degrees: Women at Oxford and the Somerville College Novelists*. New Brunswick: Rutgers University Press, 1989.

Leventhal, F. M. *The Last Dissenter: H. N. Brailsford and His World*. Oxford: Clarendon Press, 1985.

Levine, George. *Darwin and the Novelists: Patterns of Science in Victorian Fiction*. Cambridge, MA: Harvard University Press, 1988.

Lightman, Bernard. *Victorian Popularizers of Science: Designing Nature for New Audiences*. Chicago: University of Chicago Press, 2007.

Livingston, Dorothy Michelson. *The Master of Light: A Biography of Albert A. Michelson*. New York: Scribner, 1973.

Lodge, Oliver. "The New Outlook in Physics." *Discovery* 10, no. 112 (April 1929): 109–12.

Loevlie, Elisabeth Marie. *Literary Silences in Pascal, Rousseau, and Beckett*. Oxford: Clarendon Press, 2003.

Low, A. M. "Can Science Explain Love?" *Armchair Science* 1, no. 11 (February 1930): 648–51.

———. "There's Relativity in Everything." *Armchair Science* 3, no. 11 (February 1932): 646–48.

Luckhurst, Roger. *The Invention of Telepathy*. Oxford: Oxford University Press, 2002.

Macaulay, Rose. *Potterism: A Tragi-Farcical Tract*. Brussels: Collins, 1921.

Macmurray, John. Review of *The Nature of the Physical World*, by Arthur Eddington. *Criterion* 8, no. 33 (July 1929): 706–9.

Mallik, D. N. "Relativity of Time and Space." *Conquest* 2, no. 4 (February 1921): 182–85.

Mao, Douglas, and Rebecca L. Walkowitz, eds. *Bad Modernisms*. Durham: Duke University Press, 2006.

Matthews, Steven. *Modernism*. London: Arnold, 2004.

Mayer, Anna-Katherina. "Moralizing Science: The Uses of Science's Past in National Education in the 1920s." *British Journal for the History of Science* 30, no. 1 (1997): 51–70.

———. 2005. "Reluctant Technocrats: Science Promotion in the Neglect-of-Science Debate of 1916–1918." *History of Science* 43, no 140 (2005): 139–59.

McGregor, Robert Kuhn, and Ethan Lewis. *Conundrums for the Long Week-End: England, Dorothy L. Sayers, and Lord Peter Wimsey*. Kent, OH: Kent State University Press, 2000.

[Middleton Murry, John] "The Mysticism of Modern Science." *New Adelphi* 2, no. 4 (June–August 1929): 289–94.

Miller, Arthur. *Empire of the Stars: Friendship, Obsession and Betrayal in the Quest for Black Holes*. London: Little, Brown, 2005.

Moore-Brabazon, J. T. C., "Physics and the Ordinary Man." *Armchair Science* 2, no. 5 (August 1930): 289–90.

———. "Some Thoughts on Current Physics." *Armchair Science* 2, no. 11 (February 1931): 622–23.

Morrison, E. W. "The Bell That Never Rang." *Novel Magazine* 38 (February 1924): 411–14.

———. "A Wireless Uncle." *Novel Magazine* 39 (May 1924): 117–20.

Morrisson, Mark. *The Public Face of Modernism: Little Magazines, Audiences and Reception, 1905–1920*. Madison: University of Wisconsin Press, 2001.

Morus, Iwan Rhys. *When Physics Became King*. Chicago: University of Chicago Press, 2005.

Mullan, John. *How Novels Work*. Oxford: Oxford University Press, 2006.

Nevill, E., Letter to the editor. *Times*, November 17, 1919, 8.

Noakes, Richard. "'The Bridge Which Is between Physical and Psychical Research': William Fletcher Barrett, Sensitive Flames and Spiritualism." *History of Science* 42 (2004): 419–64.

———. "Telegraphy Is an Occult Art: Cromwell Fleetwood Varley and the Diffusion of Electricity to the Other World." *British Journal for the History of Science* 32, no. 4 (1999): 421–59.

———. "The 'world of the infinitely little': Connecting Physical and Psychical Realities circa 1900." *Studies in the History and Philosophy of Science*, 39, no. 3 (2008): 323–33.

Nye, Edwin R., and Mary E. Gibson. *Ronald Ross: Malariologist and Polymath*. Basingstoke: Macmillan, 1997.

O'Connor, Ralph. "Reflections on Popular Science in Britain: Genres, Categories, and Historians." *Isis* 100 (2009): 333–45.

Oppenheim, Janet. *The Other World: Spiritualism and Psychical Research in England, 1850–1914*. Cambridge: Cambridge University Press, 1985.

Otis, Laura. "The Other End of the Wire: Uncertainties of Organic and Telegraphic Communication." *Configurations* 9 (2001): 181–206.

Overton, Grant M. *Women Who Make Our Novels*. Freeport, NY: Books for Libraries, 1967.

Palladino, Paolo. "Stereochemistry and the Nature of Life: Mechanist, Vitalist and Evolutionary Perspectives." *Isis* 81 (1990): 44–67.

Pandora, Katharine. "Popular Science in National and Transnational Perspective: Suggestions from the American Context." *Isis* 100 (2009): 346–58.

Pang, Alex Soojung-Kim. *Empire and the Sun: Victorian Solar Eclipse Expeditions*. Stanford: Stanford University Press, 2002.

Parkinson, Gavin. *Surrealism, Art, and Modern Science: Relativity, Quantum Mechanics, Epistemology*. New Haven: Yale University Press, 2008.

Pascal, Blaise. *Pensées*. Paris: G. Crès, 1924. First published 1670.

Payne, F. W. *John Donne and His Poetry*. London: Harrap: 1926.

Plain, Gill. *Women's Fiction of the Second World War: Gender, Power and Resistance*. Edinburgh: Edinburgh University Press, 1996.

Prerau, David. *Saving the Daylight: Why We Put the Clocks Forward*. London: Granta, 2005.

Price, Katy. "Finite But Unbounded: *Experiment* magazine, Cambridge, England, 1928–31." *Jacket* 20 (2002). Accessed April 30, 2011. http://jacketmagazine.com/20/price-expe.html.

———. "Flame far too hot: William Empson's non-Euclidean predicament." *Interdisciplinary Science Reviews* 30, no. 4 (2005): 312–22.

———. "Monogamy and the Next Step? William Empson and the Future of Love in Einstein's Universe." In *Some Versions of Empson*, edited by Matthew Bevis, 242–63. Oxford: Oxford University Press, 2007.

———. "'On the Back of the Light Waves': Novel Possibilities in the Fourth Dimension." In *Literature and Science*, edited by Sharon Ruston, 91–110. Cambridge: Brewer, 2008.

———. "William Empson, Ants and Aliens." In *Science and Modern Poetry: New Directions*, ed. John Robert Holmes. Liverpool: Liverpool University Press, forthcoming.

Pullin, Victor. Review of *The Nature of the Physical World*, by Arthur Eddington. *Discovery* 10, no. 111 (March 1929): 104.

Pyenson, Lewis, Sean Johnston, Alberto Martínez, and Richard Staley. "Revisiting the History of Relativity." *Metascience* 20 (2011): 53–73.

Rainey, Lawrence. *Institutions of Modernism: Literary Elites and Public Cultures*. New Haven: Yale University Press, 1998.

Reeves, Barbara J. "Einstein Politicized: The Early Reception of Relativity in Italy." In *The Comparative Reception of Relativity*, edited by Thomas F. Glick, 189–229. Dordrecht: Reidel, 1987.

Reynolds, Barbara. *Dorothy L. Sayers: Her Life and Soul*. London: Hodder & Stoughton, 1993.

Reynolds, K. D. "Horner, Frances Jane, Lady Horner (1854/5–1940)." In *Oxford Dictionary of National Biography*. Oxford University Press, 2004. Accessed April 30, 2011. http://www.oxforddnb.com/view/article/49524.

Ridley, Matt. *Genome: The Autobiography of a Species in 23 Chapters*. London: Fourth Estate, 1999.

Rose, Jonathan. "Was Capitalism Good for Victorian Literature?" *Victorian Studies* 46 (2004): 489–501.

Ross, Ronald. "Mr. Lloyd George, the Nation of Shopkeepers, and the Pied Piper of Hamelin." *Science Progress* 10, no. 38 (October 1915): 315–22.

———. "Mr. Man-in-the-Mass." *Science Progress* 10, no. 39 (January 1916): 484–87.

———. "A Scientific Romance." *Science Progress* 18, no. 70 (October 1923): 279.

Rotella, Guy. "Comparing Conceptions: Frost and Eddington, Heisenberg and Bohr." *American Literature* 59 (1987): 167–90.

[Russell, A. S.] "Editorial Notes." *Discovery* 1, no. 1 (January 1920): 3–4.

———. "Editorial Notes." *Discovery* 1, no. 3 (March 1920): 67–68.

———. "Editorial Notes." *Discovery* 1, no. 10 (October 1920): 291–93.

———. "A Popular Exposition of Einstein's Theory." *Discovery* 3, no. 32 (August 1922): 220–21.

Russell, Bertrand. *ABC of Relativity*. London: Kegan Paul, 1925.

———. "The ABC of Relativity—1. Why Clocks and Footrules Mislead." *New Leader*, March 6, 1925, 12.

———. "The ABC of Relativity—2. How Space and Time Are One." *New Leader*, March 13, 1925, 10.

———. "The ABC of Relativity—3. The Eel and the Measuring Rod." *New Leader*, March 20, 1925, 12.

———. "The ABC of Relativity—4. Nature, the Anarchist." *New Leader*, March 27, 1925, 12.

———. *The Scientific Outlook*. London: George Allen & Unwin, 1931.

Ryckman, Tom. *The Reign of Relativity: Philosophy in Physics, 1915–1925*. New York: Oxford University Press, 2005.

Sainsbury, Geoffrey. "The Nature of the Physical World." *New Adelphi* 2, no. 4 (June–August 1929): 355–59.

Samuel, Herbert. "Cause, Effect, and Professor Eddington." *Nineteenth Century and After* 113 (April 1933): 469–78.

Sandage, Alan. *Centennial History of the Carnegie Institution of Washington*. Vol. 1, *The Mount Wilson Observatory*. Cambridge: Cambridge University Press, 2004.

Sayers, Dorothy L. "Absolutely Elsewhere." *Strand Magazine*, February 1934, 185–96.

———. *The Documents in the Case*. London: New English Library, 1973. First published 1930.

———. *Further Papers on Dante*. London: Methuen, 1957.

———. *Hangman's Holiday*. London: New English Library, 1982. First published 1933.

———. "Impossible Alibi." *Mystery: The Illustrated Detective Magazine*, January 1934, 19–21, 104, 106, 108.

———. *The Letters of Dorothy L. Sayers*. Edited by Barbara Reynolds. Cambridge: Dorothy L. Sayers Society, 1998.

Schaffer, Simon. "Where Experiments End: Tabletop Trials in Victorian Astronomy." In *Scientific Practice: Theories and Stories of Doing Physics*, edited by Jed Z. Buchwald, 257–99. Chicago: University of Chicago Press, 1995.

Sclove, Richard E. "From Alchemy to Atomic War: Frederick Soddy's 'Technology Assessment' of Atomic Energy, 1900–1915." *Science, Technology and Human Values* 14, no. 2 (1989): 163–94.

Secord, James A. *Victorian Sensation: The Extraordinary Publication, Reception, and Secret Authorship of* Vestiges of the Natural History of Creation. Chicago: University of Chicago Press, 2000.

Segrue, J. C., "Einstein at Home." *Daily News*, June 9, 1921, 4.

Seymour-Ure, Colin. "Northcliffe's Legacy." In *Northcliffe's Legacy: Aspects of the British Popular Press, 1896–1996*, edited by Peter Catterall, Colin Seymour-Ure, and Adrian Smith. Basingstoke: Macmillan, 2000.

Shakespeare, William. *Hamlet*. Edited by Roma Gill. Oxford: Oxford University Press, 2002.

Sieveking, L. de G. "The Lost Omnibus." *Hutchinson's Mystery-Story Magazine* 3 (April 1924): 3–6.

Sinclair, May. "The Finding of the Absolute." In *Uncanny Stories*, 225–47. London: Hutchinson, 1923.

Slater, Noel B. *The Development and Meaning of Eddington's "Fundamental Theory."* Cambridge: Cambridge University Press, 1957.

Slosson, Edwin E. *Easy Lessons in Einstein*. London: Routledge, 1920.

Spencer Jones, H. "Einstein on His Theory." *Science Progress* 14, no. 55 (January 1920): 452.

———. "Gravitation and Light." *Discovery* 1, no. 2 (February 1920): 48–51.

———. "How Soon to the Moon?" *New Scientist* 2, no. 47 (October 10 1957): 7–8.

———. "Sir Oliver Lodge and Einstein's Theory." *Science Progress* 14, no. 55 (January 1920): 451–52.

Spencer Toy, H. "Life on Other Worlds." *Conquest* 4, no. 6 (April 1923): 224–27.

Sponsel, Alistair. "Constructing a 'Revolution in Science': The Campaign to Promote a Favourable Reception for the 1919 Solar Eclipse Expeditions." *British Journal for the History of Science* 35, no. 4 (2002): 439–67.

Stachel, John. "The Theory of Relativity." In R. C. Olby, G. N. Cantor, J. R. R. Christie, and M. J. S. Hodge eds., *Companion to the History of Modern Science*, 442–56. London: Routledge, 1990.

Staley, Richard. *Einstein's Generation: The Origins of the Relativity Revolution*. Chicago: University of Chicago Press, 2008.

Stanley, Matthew. "Mysticism and Marxism: A. S. Eddington, Chapman Cohen, and Political Engagement Through Science Popularization." *Minerva: A Review of Science, Learning and Policy* 46 no. 2 (2008): 181–94.

———. *Practical Mystic: Religion, Science and A. S. Eddington*. Chicago: University of Chicago Press, 2007.

Stebbing, L. S. *Philosophy and the Physicists*. Harmondsworth: Penguin, 1944. First published 1937.

Steinman, Lisa. *Made in America: Science, Technology and American Modernist Poets*. New Haven: Yale University Press, 1987.

Stevenson, Randall. *Modernist Fiction: An Introduction*. Revised ed. Harlow: Prentice Hall, 1998. First published 1992.

Stewart, Victoria. "J. W. Dunne and Literary Culture in the 1930s and 1940s." *Literature and History* 17 (2008): 62–81.

Strutt, Robert J. (Lord Rayleigh). *The Life of Sir J. J. Thomson*. Cambridge: Cambridge University Press, 1942.

Tate, Trudi. *Modernism, History and the First World War*. Manchester: Manchester University Press, 1998.

Thompson, C. Patrick. "The Einstein Year." *Time and Tide*, June 24, 1921, 599–600.

Thompson, D'Arcy Wentworth. *On Growth and Form*. Abridged and edited by John Tyler Bonner. Cambridge: Cambridge University Press, 1992. First published 1917.

Thompson, J. Lee. *Northcliffe: Press Baron in Politics, 1865–1922*. London: John Murray, 2000.

Thomson, J. J. *Recollections and Reflections*. London: Bell, 1936.

Thorne, K. S. *Black Holes and Time Warps: Einstein's Outrageous Legacy*. New York: Norton, 1994.

———. "The Search for Black Holes." *Scientific American* 231, no. 6 (December 1974): 32–43.

Throesch, Elizabeth. "The Alice Books and the Fourth Dimension: Victorian Fantastic Spaces." In *Alice Beyond Wonderland: Essays for the Twenty-First Century*, edited by Cristopher Hollingsworth, 37–52. Iowa City: University of Iowa Press, 2009.

———. "Charles Howard Hinton's Fourth Dimension and the Phenomenology of the Scientific Romances (1884–1886)." *Foundation: The International Review of Science Fiction* 99 (2007): 29–49.

Tiffany, Daniel. *Toy Medium: Materialism and the Modern Lyric*. Berkeley: University of California Press: 2000.

Topham, Jonathan R. "Introduction: Focus on Historicizing 'Popular Science.'" *Isis* 100 (2009): 310–18.

[Turner, H. H.] "From an Oxford Note-Book." *Observatory* no. 502 (July 1916): 323.

———. "From an Oxford Note-Book." *Observatory* 42, no. 546 (December 1919): 454.

Vibert Douglas, A. *The Life of Arthur Stanley Eddington*. London: Nelson, 1956.

———. "Measuring the Universe." *Discovery* 5, no. 57 (September 1924): 196–98.

von Laue, Max. "Inertia and Energy." In *Albert Einstein, Philosopher-Scientist*, edited by P. A. Schilpp, 501–33. Evanston: Library of Living Philosophers, 1949.

Wali, K. C. *Chandra: A Biography of S. Chandrasekhar*. Chicago: University of Chicago Press, 1991.

Warwick, Andrew. "Cambridge Mathematics and Cavendish Physics: Cunningham, Campbell, and Einstein's Relativity, 1905–1911. Part 1: The Uses of Theory." *Studies in History and Philosophy of Science* 23 (1992): 625–56.

———. "Cambridge Mathematics and Cavendish Physics: Cunningham, Campbell, and Einstein's Relativity, 1905–1911. Part 2: Comparing Traditions in Cambridge Physics." *Studies in History and Philosophy of Science* 24 (1993): 1–25.

———. *Masters of Theory: Cambridge and the Rise of Mathematical Physics*. Chicago: University of Chicago Press, 2003.

Wells, H. G. *The Sleeper Awakes*. London: Collins, 1954. First published as *When the Sleeper Wakes*, 1899, revised 1910.

———. *The Time Machine*. Edited by John Lawton. Everyman New Centennial edition. London: Dent, 1995. First published 1895.

White, Eric. "Advertising Localist Modernism: William Carlos Williams, 'Aladdin Einstein' and the Transatlantic Avant-Garde in *Contact*." *European Journal of American Culture* 28 (2009): 141–65.

Whitworth, Michael. "The Clothbound Universe: Popular Physics Books, 1919–1939." *Publishing History* 40 (1996): 53–82.

———. *Einstein's Wake: Relativity, Metaphor, and Modernist Literature*. Oxford: Oxford University Press, 2001.

———. "The Physical Sciences." In *A Companion to Modernist Literature and Culture*, ed. David Bradshaw and Kevin J. H. Dettmar, 39–49. Oxford: Blackwell, 2006.

———. "Physics: 'A strange footprint.'" In *A Concise Companion to Modernism*, ed. David Bradshaw, 200–220. Oxford, Blackwell, 2003.

———. "'Pièces d'identité': T. S. Eliot, J. W. N. Sullivan and Poetic Impersonality." *English Literature in Transition* 39, no. 2 (1996): 149–70.

Wignall, T. C., and G. D. Knox. "Atoms." *Yellow Magazine* 5, nos. 29–34 (October 20–December 29, 1922).

Wilson, Leigh. "May Sinclair." *The Literary Encyclopedia*. July 17, 2001. Accessed October 14, 2009. http://www.litencyc.com/php/speople.php?rec=true&UID=4086.

Wodehouse, P. G. "The Amazing Hat Mystery." *Strand Magazine*, June 1934, 562–73.

———. "Archibald and the Masses." *Strand Magazine*, February 1936, 386–97.

———. *Mulliner Nights*. London: Herbert Jenkins, 1933.

Wootton, Barbara. "Our Autocratic Bankers." *New Leader*, March 13, 1925, 9.

Yaffle. "It Occurs to Me." *New Leader*, March 27, 1925, 20.

Ziadat, Adel. "Early Reception of Einstein's relativity in the Arab Periodical Press." *Annals of Science* 51 (1994): 17–35.

Manuscripts

Eddington Papers, Wren Library, Trinity College, University of Cambridge.

Empson Papers, MS Eng 1401, Houghton Library, Harvard University.

Index